Student Solutions Manual and Study Guide

for
Serway and Jewett's

Principles of Physics
A Calculus-Based Text

Fourth Edition
Volume 1

John R. Gordon
James Madison University

Ralph McGrew
State University of New York

Raymond A. Serway
Emeritus, James Madison University

John W. Jewett, Jr.
California State Polytechnic University – Pomona

THOMSON

BROOKS/COLE

Australia • Canada • Mexico • Singapore • Spain • United Kingdom • United States

Printer: Thomson/West

0-534-49145-6

For more information about our products,
contact us at:
Thomson Learning Academic Resource Center
1-800-423-0563

For permission to use material from this text or product, submit a request online at
http://www.thomsonrights.com.
Any additional questions about permissions can be submitted by email to **thomsonrights@thomson.com.**

Thomson Higher Education
10 Davis Drive
Belmont, CA 94002-3098
USA

Asia (including India)
Thomson Learning
5 Shenton Way
#01-01 UIC Building
Singapore 068808

Australia/New Zealand
Thomson Learning Australia
102 Dodds Street
Southbank, Victoria 3006
Australia

Canada
Thomson Nelson
1120 Birchmount Road
Toronto, Ontario M1K 5G4
Canada

UK/Europe/Middle East/Africa
Thomson Learning
High Holborn House
50–51 Bedford Road
London WC1R 4LR
United Kingdom

Latin America
Thomson Learning
Seneca, 53
Colonia Polanco
11560 Mexico
D.F. Mexico

Spain (including Portugal)
Thomson Paraninfo
Calle Magallanes, 25
28015 Madrid, Spain

CONTENTS

PREFACE

This *Student Solutions Manual and Study Guide* has been written to accompany the textbook **Principles of Physics**, Third Edition, Volume I, by Raymond A. Serway and John W. Jewett, Jr. The purpose of this Student Solutions Manual and Study Guide is to provide students with a convenient review of the basic concepts and applications presented in the textbook, together with solutions to selected end-of-chapter problems. This is not an attempt to rewrite the textbook in a condensed fashion. Rather, emphasis is placed upon clarifying typical troublesome points, and providing further practice in methods of problem solving.

Every textbook chapter has matching chapter in this book and each chapter is divided into several parts. Very often, reference is made to specific equations or figures in the textbook. Each feature of this Study Guide has been included to ensure that it serves as a useful supplement to the textbook. Most chapters contain the following components:

- **Notes From Selected Chapter Sections:** This is a summary of important concepts, newly defined physical quantities, and rules governing their behavior.

- **Equations and Concepts:** This is a review of the chapter, with emphasis on highlighting important concepts and describing important equations and formalisms.

- **Suggestions, Skills, and Strategies:** This offers hints and strategies for solving typical problems that the student will often encounter in the course. In some sections, suggestions are made concerning mathematical skills that are necessary in the analysis of problems.

- **Review Checklist:** This is a list of topics and techniques the student should master after reading the chapter and working the assigned problems.

- **Answers to Selected Conceptual Questions:** Suggested responses are provided for twenty percent of the Conceptual Questions.

- **Solutions to Selected End-of-Chapter Problems:** Solutions are given for about twelve of the odd-numbered problems in each textbook chapter. Problems were selected to illustrate important concepts in each chapter.

- **Tables:** A list of selected Physical Constants is printed on the inside front cover; and a table of some Conversion Factors is provided on the inside back cover.

An important note concerning significant figures: The answers to all end-of-chapter problems are stated to three significant figures, though calculations are carried out with as many digits as possible. We sincerely hope that this *Student Solutions Manual and Study Guide* will be useful to you in reviewing the material presented in the text, and in improving your ability to solve problems and score well on exams. We welcome any comments or suggestions that could help improve the content of this study guide in future editions; and we wish you success in your study.

John R. Gordon
Harrisonburg, VA

Ralph McGrew
Binghamton, NY

Raymond A. Serway
Leesburg, VA

John W. Jewett, Jr
Pomona, CA

ACKNOWLEDGMENTS

We take this opportunity to thank everyone who contributed to this Fourth Edition of Student Solutions Manual and Study Guide to Accompany Principles of Physics.

Special thanks for managing and directing this project go to Assistant Editor for the Physical Sciences Sarah Lowe, Development Editor for the Physical Sciences Jay Campbell, Acquisitions Editor for Physics Chris Hall, Senior Developmental Editor for Physics Susan Pashos, and Publisher for the Physical Sciences David Harris.

Our appreciation goes to our reviewer, Richard Miers of Indiana Purdue Fort Wayne. His careful reading of the manuscript and checking the accuracy of the problem solutions contributed in an important way to the quality of the final product. Any errors remaining in the manual are the responsibility of the authors.

It is a pleasure to acknowledge the excellent work of the staff of M and N Toscano who prepared the final page layout and camera-ready copy. Their technical skills and attention to detail added much to the appearance and usefulness of this volume.

Finally, we express our appreciation to our families for their inspiration, patience, and encouragement.

SUGGESTIONS FOR STUDY

We have seen a lot of successful physics students. The question, "How should I study this subject?" has no single answer, but we offer some suggestions that may be useful to you.

1. Work to understand the basic concepts and principles before attempting to solve assigned problems. Carefully read the textbook before attending your lecture on that material. Jot down points that are not clear to you, take careful notes in class, and ask questions. Reduce memorization to a minimum. Memorizing sections of a text or derivations does not necessarily mean you understand the material.

2. After reading a chapter, you should be able to define any new quantities that were introduced and discuss the first principles that were used to derive fundamental equations. A review is provided in each chapter of the Study Guide for this purpose, and the marginal notes in the textbook (or the index) will help you locate these topics. You should be able to correctly associate with each *physical quantity* the *symbol* used to represent that quantity (including vector notation, if appropriate) and the SI *unit* in which the quantity is specified. Furthermore, you should be able to express each important principle or equation in a concise and accurate prose statement. Perhaps the best test of your understanding of the material will be your ability to answer questions and solve problems in the text, or those given on exams.

3. Try to solve plenty of the problems at the end of the chapter. The worked examples in the text will serve as a basis for your study. This Study Guide contains detailed solutions to about twelve of the problems at the end of each chapter. You will be able to check the accuracy of your calculations for any odd-numbered problem, since the answers to these are given at the back of the text.

4.   Besides what you might expect to learn about physics concepts, a very valuable skill you can take away from your physics course is the ability to solve complicated problems. The way physicists approach complex situations and break them down into manageable pieces is widely useful. Starting in Section 1.10, the textbook develops a general problem-solving strategy that guides you through the steps. To help you remember the steps of the strategy, they are called *Conceptualize*, *Categorize*, *Analyze*, and *Finalize*.

GENERAL PROBLEM-SOLVING STRATEGY

Conceptualize

- The first thing to do when approaching a problem is to *think about* and *understand* the situation. Read the problem several times until you are confident you understand what is being asked. Study carefully any diagrams, graphs, tables, or photographs that accompany the problem. Imagine a movie, running in your mind, of what happens in the problem.

- If a diagram is not provided, you should almost always make a quick drawing of the situation. Indicate any known values, perhaps in a table or directly on your sketch.

- Now focus on what algebraic or numerical information is given in the problem. In the problem statement, look for key phrases such as "starts from at rest" ($v_i = 0$), "stops" ($v_f = 0$), or "freely falls" ($a_y = -g = -9.80 \text{ m/s}^2$). Key words can help simplify the problem.

- Next focus on the expected result of solving the problem. Exactly what is the question asking? Will the final result be numerical or algebraic? If it is numerical, what units will it have? If it is algebraic, what symbols will appear in it?

- Incorporate information from your own experiences and common sense. What should a reasonable answer look like? What should its order of magnitude be? You wouldn't expect to calculate the speed of an automobile to be 5×10^6 m/s.

Categorize

- Once you have a really good idea of what the problem is about, you need to *simplify* the problem. Remove the details that are not important to the solution. For example, you can often model a moving object as a particle. Key words should tell you whether you can ignore air resistance or friction between a sliding object and a surface.

- Once the problem is simplified, it is important to *categorize* the problem. How does it fit into a framework of ideas that you construct to understand the world? Is it a simple *plug-in problem*, such that numbers can be simply substituted into a definition? If so, the problem is likely to be finished when this substitution is done. If not, you face what we can call an *analysis problem*—the situation must be analyzed more deeply to reach a solution.

- If it is an analysis problem, it needs to be categorized further. Have you seen this type of problem before? Does it fall into the growing list of types of problems that you have solved previously? Being able to classify a problem can make it much easier to lay out a plan to solve it. For example, if your simplification shows that the problem can be treated as a particle moving under constant acceleration and you have already solved such a problem (such as the examples in Section 2.6), the solution to the new problem follows a similar pattern.

Analyze

- Now, you need to analyze the problem and strive for a mathematical solution. Because you have already categorized the problem, it should not be too difficult to select relevant equations that apply to the type of situation in the problem. For example, if your categorization shows that the problem involves a particle moving under constant acceleration, Equations 2.9 to 2.13 are relevant.

- Use algebra (and calculus, if necessary) to solve symbolically for the unknown variable in terms of what is given. Substitute in the appropriate numbers, calculate the result, and round it to the proper number of significant figures.

Finalize

- This final step is the most important part. Examine your numerical answer. Does it have the correct units? Does it meet your expectations from your conceptualization of the problem? What about the algebraic form of the result—before you substituted numerical values? Does it make sense? Try looking at the variables in it to see whether the answer would change in a physically meaningful way if they were drastically increased or decreased or even became zero. Looking at limiting cases to see whether they yield expected values is a very useful way to make sure that you are obtaining reasonable results.

- Think about how this problem compares with others you have done. How was it similar? In what critical ways did it differ? Why was this problem assigned? You should have learned something by doing it. Can you figure out what? Can you use your solution to expand, strengthen, or otherwise improve your framework of ideas? If it is a new category of problem, be sure you understand it so that you can use it as a model for solving future problems in the same category.

When solving complex problems, you may need to identify a series of subproblems and apply the problem-solving strategy to each. For very simple problems, you probably don't need this whole strategy. But when you are looking at a problem and you don't know what to do next, remember the steps in the strategy and use them as a guide.

Work on problems in this Study Guide yourself and compare your solutions with ours. Your solution does not have to look just like the one presented here. A problem can sometimes be solved in different ways, starting from different principles. If you wonder about the validity of an alternative approach, ask your instructor.

5.   We suggest that you use this Study Guide to review the material covered in the text, and as a guide in preparing for exams. You can use the sections Chapter Review, Notes From Selected Chapter Sections, and Equations and Concepts to focus in on any points which require further study. The main purpose of this Study Guide is to improve upon the efficiency and effectiveness of your study hours and your overall understanding of physical concepts. However, it should not be regarded as a substitute for your textbook or for individual study and practice in problem solving.

Chapter 1

Introduction and Vectors

NOTES FROM SELECTED CHAPTER SECTIONS

Section 1.1   Standards of Length, Mass, and Time

Mechanical quantities can be expressed in terms of three fundamental quantities, **mass, length,** and **time**, which in the SI system have the units **kilograms** (kg), **meters** (m), and **seconds** (s), respectively.

One **mole** (mol) of an element (or compound) is that quantity of the material that contains Avogadro's number of atoms (or molecules). **Avogadro's number** (N_A), is 6.02×10^{23}. The mass of one mole of an element, in grams, is the mass of one atom in atomic mass units.

Section 1.3   Dimensional Analysis

Dimensional analysis makes use of the fact that **dimensions can be treated as algebraic quantities**. That is, quantities can be added or subtracted only if they have the same dimensions. Furthermore, the terms on both sides of an equation must have the same dimensions. *By following these simple rules, you can use dimensional analysis to help determine whether or not an expression has the correct form, because the relationship can be correct only if the dimensions on the two sides of the equation are the same.*

Section 1.5   Significant Figures

When one performs measurements on certain quantities, the accuracy of the measured values can vary; that is, the true values are known only to be within the limits of the experimental uncertainty. The value of the uncertainty can depend on various factors such as the quality of the apparatus, the skill of the experimenter, and the number of measurements performed.

When **multiplying several quantities**, the number of significant figures in the final answer is the same as the number of significant figures in the least accurate of the quantities being multiplied. In this context "least accurate" means "having the lowest number of significant figures." The **same rule applies to division**.

When **numbers are added (or subtracted)**, the number of decimal places in the result should equal the smallest number of decimal places of any term in the sum.

Section 1.6   Coordinate Systems

The location of any point in space can be specified by the use of a coordinate system. Two frequently used coordinate systems are the **Cartesian** (or **rectangular**) coordinate system and the **plane polar coordinate** system.

In general, any coordinate system consists of a:

- fixed reference point O, called the origin;

- set of specified axes or directions with an appropriate scale and axis labels;

- set of instructions that tell us how to label a point in space relative to the origin and axes.

Section 1.7   Vectors and Scalars

A **vector** is a physical quantity that must be specified by both magnitude and direction. A **scalar** quantity has only magnitude.

Section 1.8   Some Properties of Vectors

- **Equality of Two Vectors**—Two vectors are equal vectors if they have the same magnitude and the same direction. *It is not necessary that they act along the same line.*

- **Addition of Vectors**—In order for two or more vectors to be added, they must have the same units; their sum is independent of the order of addition. The triangle method and the parallelogram method are graphical methods for determining the resultant or sum of two or more vectors.

- **Negative of a Vector**—The sum of a vector and its negative is zero. A vector and its negative have the same magnitude, but have opposite directions.

- **Multiplication by a Scalar**—When a vector is multiplied or divided by a positive (negative) scalar, the result is a vector in the same (opposite) direction. The magnitude of the resulting vector is equal to the product of the absolute value of the scalar and the magnitude of the original vector.

Section 1.9   Components of a Vector and Unit Vectors

Any vector can be completely described by its components. A unit vector is a dimensionless vector one unit in length used to specify a given direction.

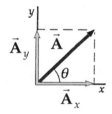

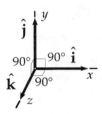

FIG. 1.1 Components of a Vector   **FIG. 1.2** Unit Vectors

The unit vectors $\hat{\mathbf{i}}$, $\hat{\mathbf{j}}$, and $\hat{\mathbf{k}}$ form a set of **mutually perpendicular** vectors as illustrated above.

$$\hat{\mathbf{i}} \equiv \text{a unit vector along the } x \text{ axis}$$
$$\hat{\mathbf{j}} \equiv \text{a unit vector along the } y \text{ axis}$$
$$\hat{\mathbf{k}} \equiv \text{a unit vector along the } z \text{ axis}$$
$$\text{where } \left|\hat{\mathbf{i}}\right| = \left|\hat{\mathbf{j}}\right| = \left|\hat{\mathbf{k}}\right| = 1$$

Mathematical Notation

It is often convenient to use symbols to represent mathematical operations and/or relationships between or among variables or values of physical quantities. Some symbols that will be used throughout your textbook and in the Solutions Manual are shown below.

Symbol	Meaning
$=$	equality of two quantities
$\approx$	approximately equal to
$\sim$	on the order of
$\equiv$	defined as
$\propto$	proportional to
$<$	less than
$>$	greater than
Δ	change in a quantity
$|x|$	magnitude of the quantity x
$\sum x_1$	sum of the set x

EQUATIONS AND CONCEPTS

The **density** of any substance is defined as the ratio of mass to volume.

$$\rho \equiv \frac{m}{V} \tag{1.1}$$

The **location of a point** P in a plane can be specified by either Cartesian coordinates, x and y, or polar coordinates, r and θ. *If one set of coordinates is known, values for the other set can be calculated.*

$$x = r\cos\theta \tag{1.2}$$

$$y = r\sin\theta \tag{1.3}$$

$$\tan\theta = \frac{y}{x} \tag{1.4}$$

$$r = \sqrt{x^2 + y^2} \tag{1.5}$$

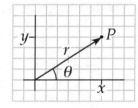

The **commutative law of addition** states that when two or more vectors are added, the sum is independent of the order of addition. In order to add vector $\vec{A}$ to vector $\vec{B}$ using the graphical method, first construct vector $\vec{A}$, and then draw vector $\vec{B}$ such that the tail of vector $\vec{B}$ starts at the head of vector $\vec{A}$. The sum of $\vec{A} + \vec{B}$ is the vector that completes the triangle by connecting the tail of $\vec{A}$ to the head of $\vec{B}$.

$$\vec{A} + \vec{B} = \vec{B} + \vec{A} \tag{1.7}$$

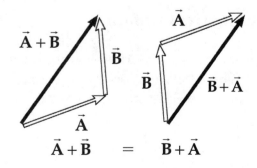

The **associative law of addition** states that when three or more vectors are added the sum is independent of the way in which the individual vectors are grouped.

$$\vec{A} + \left(\vec{B} + \vec{C}\right) = \left(\vec{A} + \vec{B}\right) + \vec{C} \qquad (1.8)$$

In the **graphical or geometric method** of vector addition, the vectors to be added (or subtracted) are represented by arrows connected head-to-tail in any order. The resultant or sum is the vector that joins the tail of the first vector to the head of the last vector. The length of each arrow must be proportional to the magnitude of the corresponding vector and must be along the direction that makes the proper angle relative to the others. *When two or more vectors are to be added, all of them must represent the same physical quantity—that is, have the same units.*

$$\vec{R} = \vec{A} + \vec{B} + \vec{C} + \vec{D}$$

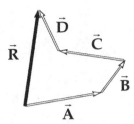

Graphical method of vector addition

The **operation of vector subtraction** utilizes the definition of the negative of a vector. The vector $(-\vec{A})$ has a magnitude equal to the magnitude of A, but acts or points along a direction opposite the direction of $\vec{A}$. *The negative of vector $\vec{A}$ is defined as the vector that when added to $\vec{A}$ gives zero for the vector sum.*

$$\vec{A} - \vec{B} = \vec{A} + \left(-\vec{B}\right) \qquad (1.9)$$

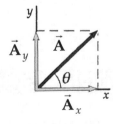

Vector subtraction

The **rectangular components** of a vector are the projections of the vector onto the respective coordinate axes. As illustrated in the figure, the projection of $\vec{A}$ onto the x axis is the x component of $\vec{A}$; and the projection of $\vec{A}$ onto the y axis is the y component of $\vec{A}$. *The angle θ is measured relative to the positive x-axis and the algebraic sign of the components will depend on the value of θ.*

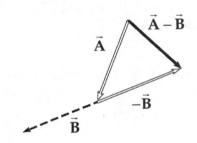

Rectangular components of a vector

The **magnitude and direction** of vector $\vec{A}$ can be determined from the values of the x and y components of $\vec{A}$.

$$A_x = A\cos\theta \tag{1.10}$$

$$A_y = A\sin\theta$$

$$A = \sqrt{A_x{}^2 + A_y{}^2} \tag{1.11}$$

$$\tan\theta = \frac{A_y}{A_x} \tag{1.12}$$

Unit vectors are dimensionless and have a magnitude of exactly 1. Vector $\vec{A}$ lying in the x-y plane, having rectangular components A_x and A_y, can be expressed in unit vector notation.

$$\vec{A} = A_x\hat{\mathbf{i}} + A_y\hat{\mathbf{j}} \tag{1.13}$$

Unit vectors specify the directions of the vector components: $\hat{\mathbf{i}}$ along the x-axis, $\hat{\mathbf{j}}$ along the y-axis and $\hat{\mathbf{k}}$ along the z-axis.

The **resultant vector** (vector sum or difference) can be expressed in terms of the components of the two vectors.

If $\qquad \vec{R} = \vec{A} + \vec{B}$, then $\tag{1.14}$

$$\vec{R} = \left(A_x + B_x\right)\hat{\mathbf{i}} + \left(A_y + B_y\right)\hat{\mathbf{j}}$$

The **components of the resultant vector** can be expressed in terms of the components of the two vectors to be added.

$$R_x = A_x + B_x$$

$$R_y = A_y + B_y \tag{1.15}$$

The **magnitude and direction** of the resultant vector can be obtained from its components.

$$R = \sqrt{R_x{}^2 + R_y{}^2} \tag{1.16}$$

$$\tan\theta = \frac{R_y}{R_x} = \frac{A_y + B_y}{A_x + B_x} \tag{1.17}$$

SUGGESTIONS, SKILLS, AND STRATEGIES

General Problem-Solving Strategy

Review carefully the discussion of forming models and alternative representations presented in Section 1.10 of your textbook. During your progress through this course you will use **simplification, analysis,** and **structural models** in solving physics problems. You will find it necessary to choose and translate among mental, pictorial, simplified pictorial, graphic, tabular, and mathematical representations as ways of viewing and presenting information.

The following basic steps are commonly used in problem-solving:

- Read the problem carefully at least twice. Be sure you understand the nature of the problem before proceeding further.

- Draw a suitable diagram with appropriate labels and coordinate axes if needed.

- Imagine a movie, running in your mind, of what happens in the problem.

- As you examine what is being asked in the problem, identify the basic physical principle or principles that are involved, listing the knowns and unknowns.

- Select a basic relationship or derive an equation that can be used to find the unknown, and symbolically solve the equation for the unknown.

- Substitute the given values, along with the appropriate units, into the equation.

- Obtain a numerical value for the unknown. The problem is verified if the following questions can be properly answered: Do the units match? Is the answer reasonable? Is the positive or negative sign proper or meaningful?

You should be familiar with some important mathematical techniques:

- Using powers of ten in expressing such numbers as $0.000\ 58 = 5.8 \times 10^{-4}$.

- Basic algebraic operations such as factoring, handling fractions, solving quadratic equations, and solving linear equations.

- The fundamentals of plane and solid geometry—including the ability to graph functions, calculate the areas and volumes of standard geometric figures and recognize the equations and graphs of a straight line, a circle, an ellipse, a parabola and a hyperbola.

- The basic ideas of trigonometry—definitions and properties of the sine, cosine, and tangent functions; the Pythagorean Theorem and some of the basic trigonometric identities.

Adding Vectors

When two or more vectors are to be added, the following step-by-step procedure is recommended:

- Select a coordinate system.

- Draw a sketch of the vectors to be added (or subtracted), with a label on each vector.

- Find the x and y components of all vectors.

- Find the resultant components (the algebraic sum of the components) in both the x and y directions.

- Use the Pythagorean theorem to find the magnitude of the resultant vector.

- Use a suitable trigonometric function to find the angle the resultant vector makes with the x axis.

REVIEW CHECKLIST

✓   Discuss the units of length, mass, and time and the standards for these quantities in SI units; and perform a dimensional analysis of an equation containing physical quantities whose individual units are known.

✓   Convert units from one system to another.

✓   Become familiar with the meaning of various mathematical symbols and Greek letters. Identify and properly use mathematical notations such as the following: $\propto$ (is proportional to), $<$ (is less than), $\approx$ (is approximately equal to), and Δ (a change in quantity).

✓   Locate a point in space using both Cartesian coordinates and polar coordinates.

✓   Describe the basic properties of vectors such as the rules of vector addition and graphical solutions for the addition of two or more vectors.

✓   Resolve a vector into its rectangular components. Determine the magnitude and direction of a vector from its rectangular components.

✓   Understand the use of unit vectors to express any vector in unit vector notation.

ANSWERS TO SELECTED QUESTIONS

Q1.2   Suppose that the three fundamental standards of the metric system were length, **density**, and time rather than length, **mass**, and time. The standard of density in this system is to be defined as that of water. What considerations about water would need to be addressed to make sure that the standard unit of density is as accurate as possible?

Answer   A number of considerations would be necessary. There are the environmental details related to the water—a standard temperature would have to be defined, as well as a standard pressure. Another consideration is the quality of the water, in terms of defining a lower limit of impurities. Another problem with this scheme is that density cannot be measured directly with a single measurement, as can length, mass, and time. As a combination of two measurements (mass, and volume, which itself involves **three** measurements!), it has higher inherent uncertainty than a single measurement.

Q1.3   Express the following quantities using the prefixes given in Table 1.4:

   (a) 3×10^{-4} m

   (b) 5×10^{-5} s

   (c) 72×10^{2} g

Answer   (a) 3×10^{-4} m $= 300 \times 10^{-6}$ m $= 300$ μm $= 0.3$ mm

   (b) 5×10^{-5} s $= 50 \times 10^{-6}$ s $= 50$ μs

   (c) 72×10^{2} g $= 7.2 \times 10^{3}$ g $= 7.2$ kg

Q1.11   A vector $\vec{A}$ lies in the xy plane. For what orientations of $\vec{A}$ will both of its components be negative? For what orientations will its components have opposite signs?

Answer   The vector $\vec{A}$ will have both rectangular components negative in quadrant III, when the angle of $\vec{A}$ is between $\pi\,\text{rad}(180°)$ and $\frac{3\pi}{2}\text{rad}(270°)$. The vector $\vec{A}$ will have components with opposite signs in two cases: first, in quadrant II, when the angle of $\vec{A}$ is between $\frac{\pi}{2}\text{rad}(90°)$ and $\pi\,\text{rad}(180°)$, and second, in quadrant IV, when the angle of $\vec{A}$ is between $\frac{3\pi}{2}\text{rad}(270°)$ and $2\pi\,\text{rad}(360°)$.

Q1.17   Is it possible to add a vector quantity to a scalar quantity? Explain.

Answer   Vectors and scalars are distinctly different and cannot be added to each other. Remember that a vector defines a quantity **in a certain direction**, while a scalar only defines a quantity with no associated direction.

SOLUTIONS TO SELECTED PROBLEMS

P1.5   The position of a particle when moving under uniform acceleration is some function of time and the acceleration. Suppose we write this position as $x = k a^m t^n$, where k is a dimensionless constant. Show by dimensional analysis that this expression is satisfied if $m = 1$ and $n = 2$. Can this analysis give the value of k?

Solution   For the equation to be valid, we must choose values of m and n to make it dimensionally consistent. Since x is a position, its dimensions are those of length (L). The acceleration, a, is a length divided by the square of a time $\left(\dfrac{L}{T^2}\right)$. The variable t has dimensions of time (T), and the constant k has no dimensions. Substituting these dimensions into the equation yields:

$$(L) = \left(\frac{L}{T^2}\right)^m (T)^n = (L)^m (T)^{-2m} (T)^n \quad \text{or} \quad (L)^1 (T)^0 = (L)^m (T)^{n-2m}.$$

continued on next page

Note that the factor $(T)^0$ introduced on the left side of the second equation is equal to 1. This equation can be true only if the powers of length (L) are the same on the two sides of the equation and, simultaneously, the powers of time (T) are the same on both sides. Indeed, if another basic unit such as mass (M) were present, we would also require that its powers be identical on the two sides. Thus, we obtain a set of two simultaneous equations:

$$1 = m \text{ and } 0 = n - 2m.$$

The solutions are therefore seen to be:

$$m = 1 \text{ and } n = 2. \qquad \lozenge$$

This gives no information about possible values of the dimensionless constant k. $\lozenge$

P1.13 One gallon of paint (volume $= 3.78 \times 10^{-3}$ m^3) covers an area of 25.0 m^2. What is the thickness of the paint on the wall?

Solution We assume the paint keeps the same volume in the can and on the wall. We model the film on the wall as a rectangular solid, with its volume given by its surface area multiplied by its uniform thickness t: $V = At$. Therefore,

$$t = \frac{V}{A} = \frac{3.78 \times 10^{-3} \text{ m}^3}{25.0 \text{ m}^2} = 1.51 \times 10^{-4} \text{ m}. \qquad \lozenge$$

P1.15 One cubic meter (1.00 m^3) of aluminum has a mass of 2.70×10^3 kg, and 1.00 m^3 of iron has a mass of 7.86×10^3 kg. Find the radius of a solid aluminum sphere that will balance a solid iron sphere of radius 2.00 cm on an equal-arm balance.

Solution We require equal masses: $\quad m_{Al} = m_{Fe} \quad$ or $\quad \rho_{Al} V_{Al} = \rho_{Fe} V_{Fe}$. Therefore,

$$\rho_{Al}\left(\frac{4}{3}\pi r^3\right) = \rho_{Fe}\left(\frac{4}{3}\pi (2.00 \text{ cm})^3\right)$$

$$r^3 = \left(\frac{\rho_{Fe}}{\rho_{Al}}\right)(2.00 \text{ cm})^3 = \left(\frac{7.86 \text{ kg/m}^3}{2.70 \text{ kg/m}^3}\right)(2.00 \text{ cm})^3 = 23.3 \text{ cm}^3$$

$$r = 2.86 \text{ cm}. \qquad \lozenge$$

P1.19   Estimate the number of Ping-Pong balls that would fit into a typical-size room (without being crushed). In your solution state the quantities you measure or estimate and the values you take for them.

Solution   **Conceptualize.** Since the volume of a typical room is much larger than a Ping-Pong ball, we should expect that a very large number of balls (maybe a million) could fit in a room.

Categorize. Since we are only asked to find an estimate, we do not need to be too concerned about how the balls are arranged. Therefore, to find the number of balls we can simply divide the volume of an average-size living room (perhaps $15 \text{ ft} \times 20 \text{ ft} \times 8 \text{ ft}$) by the volume of an individual Ping-Pong ball.

Analyze. Using the approximate conversion $1 \text{ ft} \approx 30 \text{ cm}$, we find

$$V_{\text{Room}} = (15 \text{ ft})(20 \text{ ft})(8 \text{ ft})(30 \text{ cm/ft})^3 \approx 7 \times 10^7 \text{ cm}^3.$$

A Ping-Pong ball has a diameter of about 3 cm, so we can estimate its volume as a cube:

$$V_{\text{ball}} = (3 \text{ cm})(3 \text{ cm})(3 \text{ cm}) \approx 30 \text{ cm}^3.$$

The number of Ping-Pong balls that can fill the room is

$$N \approx \frac{V_{\text{Room}}}{V_{\text{ball}}} \approx 2 \times 10^6 \text{ balls} \sim 10^6 \text{ balls}. \qquad \lozenge$$

Finalize. So a typical room can hold about a million Ping-Pong balls. This problem gives us a sense of how large a quantity "a million" really is.

P1.33   A fly lands on one wall of a room. The lower left corner of the wall is selected as the origin of a two-dimensional Cartesian coordinate system. If the fly is located at the point having coordinates (2.00, 1.00) m,

(a)   How far is it from the corner of the room?

(b)   What is its location in polar coordinates?

continued on next page

Solution (a) Assume the wall is in the *x-y* plane so that the coordinates are $x = 2.00$ m and $y = 1.00$ m; and the fly is located at point P. The distance between two points in the *x - y* plane is

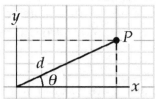

FIG. P1.33

$$d = \sqrt{(x_2 - x_1)^2 + (y_2 - y_1)^2}$$
$$d = \sqrt{(2.00 \text{ m} - 0)^2 + (1.00 \text{ m} - 0)^2} = 2.24 \text{ m}.$$

◊

(b) From the figure,

$$\theta = \tan^{-1}\left(\frac{x}{y}\right) = \tan^{-1}\left(\frac{1.00}{2.00}\right) = 26.6° \text{ and } r = d.$$

Therefore, the polar coordinates of the point P are (2.24 m, 26.6°). ◊

P1.37 A skater glides along a circular path of radius 5.00 m. Assuming he coasts around one half of the circle, find

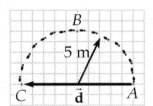

(a) the magnitude of the displacement vector.

(b) how far the person skated.

FIG. P1.37

(c) What is the magnitude of the displacement if he skates all the way around the circle?

Solution See Fig. P1.37.

(a) $\left|\vec{\mathbf{d}}\right| = \left|-10.0\hat{\mathbf{i}}\right| = 10.0$ m ◊

since the displacement is a straight line from point A to point C.

(b) The actual distance skated is not equal to the straight-line displacement. The distance follows the curved path of the semicircle (ABC).

$$s = \left(\frac{1}{2}\right)(2\pi r) = 5.00\pi = 15.7 \text{ m}$$

◊

(c) If the circle is complete, $\vec{\mathbf{d}}$ begins and ends at point A. Hence,

$$\left|\vec{\mathbf{d}}\right| = 0.$$

◊

P1.39   A roller coaster car moves 200 ft horizontally and then rises 135 ft at an angle of 30.0° above the horizontal. It then travels 135 ft at an angle of 40.0° downward. What is its displacement from its starting point? Use graphical techniques.

Solution Your sketch when drawn to scale should look somewhat like the one to the right. (You will probably only be able to obtain a measurement to one or two significant figures).

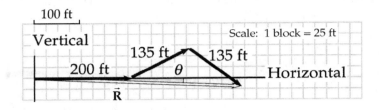

FIG. P1.39

The distance R and the angle θ can be measured to give, upon use of your scale factor, the values of:

$$\vec{R} = 4.2 \times 10^2 \text{ ft at about 3° below the horizontal.} \qquad \Diamond$$

P1.45   Consider two vectors $\vec{A} = 3\hat{i} - 2\hat{j}$ and $\vec{B} = -\hat{i} - 4\hat{j}$. Calculate

(a) $\vec{A} + \vec{B}$

(b) $\vec{A} - \vec{B}$

(c) $\left| \vec{A} + \vec{B} \right|$

(d) $\left| \vec{A} - \vec{B} \right|$

(e)   the directions of $\vec{A} + \vec{B}$ and $\vec{A} - \vec{B}$.

Solution Use the property of vector addition that states that

$$\text{if } \vec{R} = \vec{A} + \vec{B} \text{ then } R_x = A_x + B_x \text{ and } R_y = A_y + B_y.$$

(a) $\vec{A} + \vec{B} = \left(3\hat{i} - 2\hat{j}\right) + \left(-\hat{i} - 4\hat{j}\right) = 2\hat{i} - 6\hat{j}$ $\qquad \Diamond$

(b) $\vec{A} - \vec{B} = \left(3\hat{i} - 2\hat{j}\right) - \left(-\hat{i} - 4\hat{j}\right) = 4\hat{i} + 2\hat{j}$ $\qquad \Diamond$

continued on next page

(c)   For a vector, if $\vec{\mathbf{R}} = R_x\hat{\mathbf{i}} + R_y\hat{\mathbf{j}}$ $\left|\vec{\mathbf{R}}\right| = \sqrt{R_x^2 + R_y^2}$

$\left|\vec{\mathbf{A}} + \vec{\mathbf{B}}\right| = \sqrt{2^2 + (-6)^2} = 6.32$ ◊

(d) $\left|\vec{\mathbf{A}} - \vec{\mathbf{B}}\right| = \sqrt{4^2 + 2^2} = 4.47$ ◊

The direction of a vector relative to the positive x axis is given by

$\theta = \tan^{-1}\left(\dfrac{R_y}{R_x}\right)$.

(e) $\vec{\mathbf{A}} + \vec{\mathbf{B}}$ is in the fourth quadrant, with direction angle,

$\theta = \tan^{-1}\left(-\dfrac{6}{2}\right) = -71.6° = 288°$.   ◊

$\vec{\mathbf{A}} - \vec{\mathbf{B}}$ is in the first quadrant. For it, $\theta = \tan^{-1}\left(\dfrac{2}{4}\right) = 26.6°$.   ◊

P1.47   A person going for a walk follows the path shown in Figure P1.47. The total trip consists of four straight-line paths. At the end of the walk, what is the person's resultant displacement measured from the starting point?

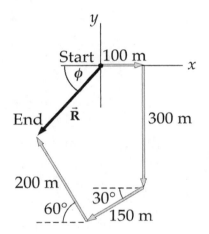

Solution   The resultant displacement $\vec{\mathbf{R}}$ is equal to the sum of the four individual displacements,

$$\vec{\mathbf{R}} = \vec{\mathbf{d}}_1 + \vec{\mathbf{d}}_2 + \vec{\mathbf{d}}_3 + \vec{\mathbf{d}}_4.$$

We translate from the pictorial representation to a mathematical representation by writing the individual displacements in unit-vector notation:

FIG. P1.47

$\vec{\mathbf{d}}_1 = 100\hat{\mathbf{i}}$ m

$\vec{\mathbf{d}}_2 = -300\hat{\mathbf{j}}$ m

$\vec{\mathbf{d}}_3 = (-150\cos 30°)\hat{\mathbf{i}}$ m $+ (-150\sin 30°)\hat{\mathbf{j}}$ m $= -130\hat{\mathbf{i}}$ m $- 75\hat{\mathbf{j}}$ m

$\vec{\mathbf{d}}_4 = (-200\cos 60°)\hat{\mathbf{i}}$ m $+ (200\sin 60°)\hat{\mathbf{j}}$ m $= -100\hat{\mathbf{i}}$ m $+ 173\hat{\mathbf{j}}$ m.

continued on next page

Summing the components together,

$$R_x = d_{1x} + d_{2x} + d_{3x} + d_{4x} = (100 + 0 - 130 - 100)\ \text{m} = -130\ \text{m}$$

$$R_y = d_{1y} + d_{2y} + d_{3y} + d_{4y} = (0 - 300 - 75 + 173)\ \text{m} = -202\ \text{m}$$

$$\vec{R} = -130\hat{i}\ \text{m} - 202\hat{j}\ \text{m} \qquad \diamond$$

$$\left|\vec{R}\right| = \sqrt{R_x^2 + R_y^2} = \sqrt{(-130\ \text{m})^2 + (-202\ \text{m})^2} = 240\ \text{m}$$

$$\phi = \tan^{-1}\left(\frac{R_y}{R_x}\right) = \tan^{-1}\left(\frac{-202}{-130}\right) = 57.2°. \qquad \diamond$$

The resultant is in the third quadrant. Pitfall Prevention 1.4 in the text explains why this angle does not specify the resultant's direction in standard form. Instead, the angle counterclockwise from the +x axis is

$$\theta = 57.2° + 180° = 237°. \qquad \diamond$$

P1.51   The vector $\vec{A}$ has x, y, and z components of 8.00, 12.0, and –4.00 units, respectively.

(a)   Write a vector expression for $\vec{A}$ in unit-vector notation.

(b)   Obtain a unit-vector expression for a vector $\vec{B}$ one-fourth the length of $\vec{A}$ pointing in the same direction as $\vec{A}$.

(c)   Obtain a unit-vector expression for a vector $\vec{C}$ three times the length of $\vec{A}$ pointing in the direction opposite the direction of $\vec{A}$.

Solution (a) $\vec{A} = A_x\hat{i} + A_y\hat{j} + A_z\hat{k}$ $\vec{A} = 8.00\hat{i} + 12.0\hat{j} - 4.00\hat{k}$ $\qquad \diamond$

(b) $\vec{B} = \dfrac{\vec{A}}{4}$ $\vec{B} = 2.00\hat{i} + 3.00\hat{j} - 1.00\hat{k}$ $\qquad \diamond$

(c) $\vec{C} = -3\vec{A}$ $\vec{C} = -24.0\hat{i} - 36.0\hat{j} + 12.0\hat{k}$ $\qquad \diamond$

P1.53 Three displacement vectors of a croquet ball are shown in Figure P1.53, where $|\vec{A}| = 20.0$ units, $|\vec{B}| = 40.0$ units, and $|\vec{C}| = 30.0$ units.

(a) Find the resultant in unit-vector notation.

(b) Find the magnitude and direction of the resultant vector.

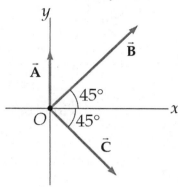

Solution Conceptualize. The given diagram shows the vectors individually, but not their addition. The second diagram represents a map view of the motion of the ball. From it, the magnitude of the resultant $\vec{R}$ should be about 60 units. Its direction is in the first quadrant, at something like 30° from the x axis. Its x component appears to be about 50 units and its y component about 30 units.

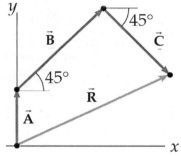

Categorize. According to the definition of a displacement, we ignore any departure of the actual path of the ball from straightness. We model each of the three motions as straight. The simplified problem is solved by straightforward application of the component method of vector addition. It works for adding two, three, or any number of vectors.

FIG. P1.53

Analyze

(a) $A_x = (20.0 \text{ units}) \cos 90° = 0$

$A_y = (20.0 \text{ units}) \sin 90° = 20.0 \text{ units}$

$B_x = (40.0 \text{ units}) \cos 45° = 28.3 \text{ units}$

$B_y = (40.0 \text{ units}) \sin 45° = 28.3 \text{ units}$

$C_x = (30.0 \text{ units}) \cos 315° = 21.2 \text{ units}$

$C_y = (30.0 \text{ units}) \sin 315° = -21.2 \text{ units}$

$R_x = A_x + B_x + C_x = (0 + 28.3 + 21.2) \text{ units} = 49.5 \text{ units}$

$R_y = A_y + B_y + C_y = (20 + 28.3 - 21.2) \text{ units} = 27.1 \text{ units}$

$\vec{R} = 49.5\hat{\mathbf{i}} + 27.1\hat{\mathbf{j}} \text{ units}$ ◊

continued on next page

(b) $\quad |\vec{R}| = \sqrt{R_x^2 + R_y^2} = \sqrt{(49.5 \text{ units})^2 + (27.1 \text{ units})^2} = 56.4 \text{ units}$ ◊

$$\theta = \tan^{-1}\left(\frac{R_y}{R_x}\right) = \tan^{-1}\left(\frac{27.1}{49.5}\right) = 28.7°$$ ◊

Finalize. The approximate values we guessed in the conceptualize step $(60 \text{ units at } 30° \approx (50\hat{i} + 30\hat{j}) \text{ units})$ agree with the computed values precise to three digits $(56.4 \text{ units at } 28.7° = (49.5\hat{i} + 27.1\hat{j}) \text{ units})$. Perhaps the greatest usefulness of the diagram is checking positive and negative signs for each component of each vector. The y component of $\vec{C}$ is negative, for example, because the vector is downward. For each vector it is good to check against the diagram whether the y component is less than, equal to, or greater than the x component. If your calculator is set to radians instead of degrees, the diagram can rescue you.

P1.61 There are nearly $\pi \times 10^7$ s in one year. Find the percentage error in this approximation, where "percentage error" is defined as

$$\text{Percentage error} = \frac{|\text{assumed value } - \text{ true value}|}{\text{true value}} \times 100\%.$$

Solution First evaluate the "true value." Remember that every fourth year is a leap year; therefore there are 365.25 days in an average year.

$$1 \text{ yr} = 1 \text{ yr}\left(\frac{365.25 \text{ d}}{1 \text{ yr}}\right)\left(\frac{24 \text{ h}}{1 \text{ d}}\right)\left(\frac{3\,600 \text{ s}}{1 \text{ h}}\right) = 3.155\,8 \times 10^7 \text{ s}$$

$$\text{Percentage error} = \frac{|\text{assumed value} - \text{true value}|}{\text{true value}} \times 100\%$$

$$\text{Percentage error} = \frac{(3.155\,8 - 3.141\,6) \times 10^7}{3.155\,8 \times 10^7} \times 100\% = 0.449\%$$ ◊

Motion in One Dimension

NOTES FROM SELECTED CHAPTER SECTIONS

Section 2.2   Instantaneous Velocity

The velocity of a particle at any instant of time (i.e. at some point on a space-time graph) is called the **instantaneous velocity**. *The slope of the line tangent to the position-time curve at a point P is defined to be the instantaneous velocity at the corresponding time.*

The **instantaneous speed** of an object, which is a scalar quantity, is defined as the magnitude of the instantaneous velocity. *Hence, by definition, speed can never be negative.*

Section 2.4   Acceleration

The **average acceleration** during a given time interval is defined as the change in velocity divided by the time interval during which this change occurs.

The **instantaneous acceleration** of an object at a certain time equals the slope of the tangent to the velocity-time graph at that instant of time.

Section 2.7   Freely Falling Objects

A freely falling body is an object moving freely under the influence of gravity only, regardless of its initial motion. *Once they are in free fall, all objects have an acceleration downward equal to the acceleration due to gravity. This is true regardless of the initial motion of the object(upward, downward, or released from rest).*

EQUATIONS AND CONCEPTS

The **displacement** Δx of a particle moving from position x_i to position x_f equals the final coordinate minus the initial coordinate. *Displacement should not be confused with distance traveled (the path length).*

$$\Delta x \equiv x_f - x_i$$

Distance traveled is the length of path followed by a particle and should not be confused with displacement. When $x_f = x_i$, the displacement is zero; however, if the particle leaves x_i, travels along a path, and returns to x_f, the distance traveled will not be zero. In the figure the distance traveled from x_i to x_f is the path length L and the displacement is the vector quantity $\vec{D}$.

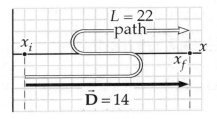

A typical path length L, and displacement $\vec{D}$ (arbitrary units)

The **average velocity** of a particle during a time interval is the ratio of the total displacement to the time interval during which the displacement occurs. Equation 2.2 is written for the case of motion along the x-axis.

$$v_{x,\,ave} \equiv \frac{\Delta v}{\Delta t} = \frac{x_f - x_i}{t_f - t_i} \qquad (2.2)$$

The **average speed** of a particle is the scalar quantity defined as the ratio of the total distance traveled to the time interval required to travel that distance. Average speed has no direction and carries no algebraic sign. *The magnitude of the average velocity is not the average speed; although in certain cases they may be numerically equal.*

$$\frac{\text{average}}{\text{speed}} \equiv \frac{\text{total distance}}{\text{total time}}$$

The **instantaneous velocity** v is defined as the limit of the ratio $\dfrac{\Delta x}{\Delta t}$ as Δt approaches zero. This limit is called the derivative of x with respect to t. The instantaneous velocity at any time is the slope of the position-time graph at that time. As illustrated in the figure, the slope can be positive, negative , or zero.

$$v_x \equiv \lim_{\Delta t \to 0} \frac{\Delta x}{\Delta t} = \frac{dx}{dt} \qquad (2.3)$$

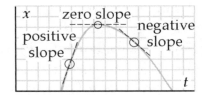

Instantaneous velocity at three points along a distance-time graph

The **instantaneous speed** is the magnitude of the instantaneous velocity.

The **average acceleration** of an object during a time interval is the ratio of the change in velocity to the time interval during which the change in velocity occurs. Equation 2.6 is for the special case of one-dimensional motion along the x-axis.

$$a_{x,\,\text{avg}} \equiv \frac{v_{xf} - v_{xi}}{t_f - t_i} = \frac{\Delta v_x}{\Delta t} \qquad (2.6)$$

The **instantaneous acceleration** a_x is defined as the limit of the ratio $\dfrac{\Delta v_x}{\Delta t}$ as Δt approaches zero.

$$a_x \equiv \lim_{\Delta t \to 0} \frac{\Delta v_x}{\Delta t} = \frac{dv_x}{dt} \qquad (2.7)$$

A negative value for acceleration does not necessarily imply that the magnitude of the velocity (speed) is deceasing.

One-dimensional motion along the x axis with constant acceleration can be described using Equations 2.9–2.13. Note that each equation shows a different relationship among physical quantities: initial velocity, final velocity, acceleration, time, and position.

$$v_{xf} = v_{xi} + a_x t \qquad (2.9)$$

$$v_{x,\,\text{avg}} = \tfrac{1}{2}\left(v_{xi} + v_{xf}\right) \qquad (2.10)$$

$$x_f = x_i + \tfrac{1}{2}(v_{xi} + v_{xf})t \qquad (2.11)$$

$$x_f = x_i + v_{xi} t + \tfrac{1}{2}a_x t^2 \qquad (2.12)$$

$$v_{xf}{}^2 = v_{xi}{}^2 + 2a_x\left(x_f - x_i\right) \qquad (2.13)$$

A **freely falling object** is any object moving under the influence of the gravitational force alone. Equations 2.9–2.13 can be modified to describe the motion of freely falling objects by denoting the motion to be along the y-axis (defining "up" as positive) and setting $a_x = -g$. *A freely falling object experiences an acceleration that is directed downward regardless of its actual motion.*

$$v_{yf} = v_{yi} - gt$$

$$y_f - y_i = \frac{1}{2}(v_{yf} + v_{yi})t$$

$$y_f - y_i = v_{yi}t - \frac{1}{2}gt^2$$

$$v_{yf}^2 = v_{yi}^2 - 2g(y_f - y_i)$$

SUGGESTIONS, SKILLS, AND STRATEGIES

Organize your problem-solving strategy by considering each step of the **Conceptualize, Categorize, Analyze,** and **Finalize** protocol described in your textbook.

Include the following steps when solving problems that involve an object undergoing constant acceleration:

- Make sure all the units in the problem are consistent. That is, if distances are measured in meters, be sure that velocities have units of m/s and accelerations have units of m/s^2.

- Choose a coordinate system.

- Choose an instant to be called the "initial point," and another to be called the "final point."

- Think about what is going on physically in the problem, and then select from the list of kinematic equations those that will enable you to determine the unknowns.

- Construct an appropriate motion diagram or a graphical representation. Check to see if your answers are consistent with the diagram.

REVIEW CHECKLIST

✓ Define the displacement and average velocity of a particle in motion. Define the instantaneous velocity and understand how this quantity differs from average velocity.

✓ Define average acceleration and instantaneous acceleration.

✓   Construct a graph of position versus time (given a function such as $x = 5 + 3t - 2t^2$) for a particle in motion along a straight line. From this graph, you should be able to determine both average and instantaneous values of velocity by calculating the slope of the tangent to the graph.

✓   Describe what is meant by a body in free fall (one moving under the influence of gravity—where air resistance is neglected). Recognize that the equations of kinematics apply directly to a freely falling object and that the acceleration is then given by $a_y = -g$ (where $g = 9.80 \text{ m/s}^2$).

✓   Apply the Equations 2.9–2.13 to any situation where the motion occurs under constant acceleration.

ANSWERS TO SELECTED QUESTIONS

Q2.3   If the average velocity of an object is zero in some time interval, what can you say about the displacement of the object for that interval?

Answer   The displacement is **zero**, since the displacement is proportional to average velocity.

Q2.9   Two cars are moving in the same direction in parallel lanes along a highway. At some instant, the velocity of car A exceeds the velocity of car B. Does this mean that the acceleration of A is greater than that of B? Explain.

Answer   No. If Car A has been traveling with cruise control, its velocity will be high (60 mph), but its acceleration will be close to zero. If Car B is pulling onto the highway, its velocity is likely to be low (30 mph), but its acceleration will be high.

Q2.11   Consider the following combinations of signs and values for velocity and acceleration of a particle with respect to a one-dimensional x axis:

	(a)	(b)	(c)	(d)	(e)	(f)	(g)	(h)
Velocity	+	+	+	−	−	−	0	0
Acceleration	+	−	0	+	−	0	+	−

Describe what a particle is doing in each case, and give a real-life example for an automobile on an east–west one-dimensional axis, with east considered the positive direction.

continued on next page

Answer   (a)   The particle is moving to the right (in the $+x$ direction) since the velocity is positive, and its speed is increasing, because the acceleration is in the same direction as the velocity. This would be the case in an automobile that is moving toward the east, starting up after waiting for a red light.

(b)   The particle is moving to the right, since the velocity is positive, and its speed is decreasing, because the acceleration is in the opposite direction to the velocity. An automobile is moving toward the east, slowing down in preparation for a red light.

(c)   The particle is moving to the right, since the velocity is positive, and its speed is constant, because the acceleration is zero. An automobile is moving toward the east, moving at constant speed on a freeway.

(d)   The particle is moving to the left (in the $-x$ direction), since the velocity is negative, and its speed is decreasing, because the acceleration is in the opposite direction to the velocity. An automobile is moving toward the west, slowing down in preparation for a red light.

(e)   The particle is moving to the left (in the $-x$ direction), since the velocity is negative, and its speed is increasing, because the acceleration is in the same direction as the velocity. An automobile is moving toward the west, starting up after waiting for a red light.

(f)   The particle is moving to the left (in the $-x$ direction), since the velocity is negative, and its speed is constant, because the acceleration is zero. An automobile is moving toward the west, moving at constant speed on a freeway.

(g)   The particle is momentarily at rest, and, in the next instant, will be moving to the right. This situation can only exist for an instant, since as soon as the particle begins to move, the velocity will no longer be zero. This is the situation for an automobile that has been coasting uphill toward the west, just when it comes to a stop before beginning to roll downhill to the east.

(h)   The particle is momentarily at rest, and, in the next instant, will be moving to the left. This situation can only exist for an instant, since as soon as the particle begins to move, the velocity will no longer be zero. This is the situation for an automobile that has been coasting uphill to the east, just when it comes to a stop before beginning to roll downhill to the west.

Q2.15   A student at the top of a building of height h throws one ball upward with a speed v_i and then throws a second ball downward with the same initial speed. How do the final velocities of the balls compare when they reach the ground?

Answer   They are the same. After the first ball reaches its apex and falls back downward past the student, it will have a downward velocity equal to v_i. This velocity is the same as the velocity of the second ball, so after they fall through equal heights their impact speeds will also be the same.

SOLUTIONS TO SELECTED PROBLEMS

P2.3   The position versus time for a certain particle moving along the x axis is shown in Figure P2.3. Find the average velocity in the time intervals.

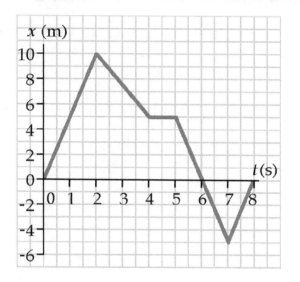

(a)   0 to 2 s

(b)   0 to 4 s

(c)   2 s to 4 s

(d)   4 s to 7 s

(e)   0 to 8 s

FIG. P2.3

Solution   On this graph, we can tell positions to two significant figures:

(a) $x = 0$ at $t = 0$ and $x = 10$ m at $t = 2$ s:

$$\overline{v}_x = \frac{\Delta x}{\Delta t} = \frac{10 \text{ m} - 0}{2 \text{ s} - 0} = 5.0 \text{ m/s} \qquad \Diamond$$

(b) $x = 5.0$ m at $t = 4$ s

$$\overline{v}_x = \frac{\Delta x}{\Delta t} = \frac{5.0 \text{ m} - 0}{4 \text{ s} - 0} = 1.2 \text{ m/s} \qquad \Diamond$$

(c) $\overline{v}_x = \frac{\Delta x}{\Delta t} = \frac{5.0 \text{ m} - 10 \text{ m}}{4 \text{ s} - 2 \text{ s}} = -2.5 \text{ m/s} \qquad \Diamond$

continued on next page

(d) $\bar{v}_x = \dfrac{\Delta x}{\Delta t} = \dfrac{-5.0 \text{ m} - 5.0 \text{ m}}{7 \text{ s} - 4 \text{ s}} = -3.3 \text{ m/s}$ ◇

(e) $\bar{v}_x = \dfrac{\Delta x}{\Delta t} = \dfrac{0.0 \text{ m} - 0.0 \text{ m}}{8 \text{ s} - 0 \text{ s}} = 0 \text{ m/s}$ ◇

P2.5   A position-time graph for a particle moving
along the *x* axis is shown in Figure P2.5.

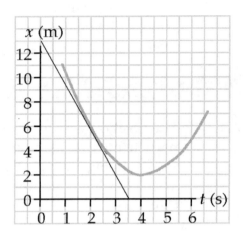

(a)   Find the average velocity in the time
interval $t = 1.50 \text{ s}$ to $t = 4.00 \text{ s}$.

(b)   Determine the instantaneous velocity
at $t = 2.00 \text{ s}$ by measuring the slope of
the tangent line shown in the graph.

(c)   At what value of *t* is the velocity zero?

FIG. P2.5

Solution (a)   From the graph:

At $t_1 = 1.5 \text{ s}$, $x = x_1 = 8.0 \text{ m}$

At $t_2 = 4.0 \text{ s}$, $x = x_2 = 2.0 \text{ m}$

Therefore,

$$\bar{v}_{1 \to 2} = \frac{\Delta x}{\Delta t} = \frac{2.0 \text{ m} - 8.0 \text{ m}}{4.0 \text{ s} - 1.5 \text{ s}} = -2.4 \text{ m/s}.$$ ◇

(b)   Choose two points along the line that is tangent to the curve at $t = 2.0 \text{ s}$.
We will use the two points:

$(t_i = 0.0 \text{ s}, x_i = 13.0 \text{ m})$ and $(t_f = 3.5 \text{ s}, x_f = 0.0 \text{ m})$.

Instantaneous velocity equals the slope of the tangent line, so

$$v = \frac{x_f - x_i}{t_f - t_i} = \frac{0.0 \text{ m} - 13.0 \text{ m}}{3.5 \text{ s} - 0.0 \text{ s}} = -3.7 \text{ m/s}.$$

The negative sign indicates that the **direction** of $\vec{v}$ is along the negative *x*
direction.   ◇

continued on next page

(c)   The velocity will be zero when the slope of the tangent line is zero. This occurs for the point on the graph where x has its minimum value. Therefore,

$$v = 0 \text{ at } t = 4.0 \text{ s.}$$ ◊

P2.13   A particle moves along the x axis according to the equation $x = 2.00 + 3.00t - 1.00t^2$, where x is in meters and t is in seconds. At $t = 3.00$ s, find

(a)   the position of the particle

(b)   its velocity

(c)   its acceleration

Solution   With the position given by $x = 2.00 + 3.00t - t^2$, we can use the rules for differentiation to write expressions for the velocity and acceleration as functions of time:

$$v = \frac{dx}{dt} = 3.00 - 2t \text{ and } a = \frac{dv}{dt} = -2.$$

Now we can evaluate x, v, and a at $t = 3.00$ s.

(a) $x = 2.00 + 3.00(3.00) - (3.00)^2 = 2.00$ m   ◊

(b) $v = 3.00 - 2(3.00) = -3.00$ m/s   ◊

(c) $a = -2.00$ m/s^2 ◊

P2.19   An object moving with uniform acceleration has a velocity of 12.0 cm/s in the positive x direction when its x coordinate is 3.00 cm. If its x coordinate 2.00 s later is –5.00 cm, what is its acceleration?

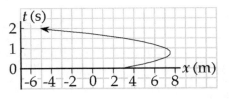

(acceleration is to the left)

FIG. P2.19

Solution   Take $t = 0$ to be the time when $x_i = 3.00$ cm.

And $v_{xi} = 12.0$ cm/s.

Also, at $t = 2.00$ s $x_f = -5.00$ cm.

continued on next page

Use the kinematic equation
$$x_f = x_i + v_{xi}t + \frac{1}{2}at^2$$

and solve for *a*.
$$a = \frac{2[x_f - x_i - v_{xi}t]}{t^2}$$

$$a = \frac{2[-5.00 \text{ cm} - 3.00 \text{ cm} - (12.0 \text{ cm/s})(2.00 \text{ s})]}{(2.00 \text{ s})^2}$$

$$a = -16.0 \text{ cm/s}^2. \qquad \Diamond$$

P2.21   A jet plane coming in for a landing with a speed of 100 m/s and can accelerate at a maximum rate of -5.00 m/s^2 as it comes to rest.

(a)   From the instant the plane touches the runway, what is the minimum time interval needed before it can come to rest?

(b)   Can this plane land at a small tropical island airport where the runway is 0.800 km long?

Solution   The negative acceleration of the plane as it lands can be called **deceleration**; however, physicists use the single term **acceleration** to describe all changes in velocity.

(a)   Assume that the acceleration of the plane is **constant** at the maximum rate, so that the plane can be modeled as a particle under constant acceleration $a_x = -5.00$ m/s^2. Given $v_{xi} = 100$ m/s and $v_{xf} = 0$, use the equation $v_{xf} = v_{xi} + a_x t$ and solve for *t*:

$$t = \frac{v_{xf} - v_{xi}}{a_x} = \frac{0 - 100 \text{ m/s}}{-5.00 \text{ m/s}^2} = 20.0 \text{ s}. \qquad \Diamond$$

(b)   Find the required stopping distance and compare this to the length of the runway. Taking x_i to be zero, we get

$$v_{xf}^2 = v_{xi}^2 + 2a_x(x_f - x_i)$$

or $\Delta x = x_f - x_i = \dfrac{v_{xf}^2 - v_{xi}^2}{2a_x} = \dfrac{0 - (100 \text{ m/s})^2}{2(-5.00 \text{ m/s}^2)} = 1\,000$ m. The stopping distance is greater than the length of the runway; the plane **cannot land.**

P2.29   A baseball is hit so that it travels straight upward after being struck by the bat. A fan observes that it takes 3.00 s for the ball to reach its maximum height.

(a)   Find its initial velocity.

(b)   Find the height it reaches.

Solution **Conceptualize.** The initial speed of the ball is probably somewhat greater than the speed of the pitch, which might be about 60 mph (~30 m/s), so an initial upward velocity off the bat of somewhere between 20 and 100 m/s would be reasonable. We also know that the length of a ball field is about 300 ft (~100 m), and a pop fly usually does not go higher than this distance, so a maximum height of 10 to 100 m would be reasonable for the situation described in this problem.

Categorize. Since the ball's motion is entirely vertical, we can use the equations for free fall to find the initial velocity and maximum height from the elapsed time.

Analyze. After leaving the bat, the ball is in free fall for $t = 3.00$ s and has a constant acceleration, $a_y = -g = -9.80 \text{ m/s}^2$.

Solve the equation $v_{yf} = v_{yi} + a_y t$ with $a_y = -g$ to obtain v_{yi} when $v_{yf} = 0$. and the ball reaches its maximum height.

(a) $v_{yi} = v_{yf} + gt = 0 + \left(9.80 \text{ m/s}^2\right)(3.00 \text{ s}) = 29.4 \text{ m/s}$ (upward)   ◊

(b)   The maximum height in the vertical direction is $y_f = v_{yi}t - \dfrac{1}{2}gt^2$:

$$y_f = (29.4 \text{ m/s})(3.00 \text{ s}) - \frac{1}{2}\left(9.80 \text{ m/s}^2\right)(3.00 \text{ s})^2 = 44.1 \text{ m}$$ ◊

Finalize: The calculated answers seem reasonable since they lie within our expected ranges, and they have the correct units and direction. We must remember that it is possible to solve a problem like this correctly, yet the answers may not seem reasonable simply because the conditions stated in the problem may not be physically possible (e.g., a time of 10 seconds for a pop fly would not be realistic).

P2.31 A student throws a set of keys vertically upward to her sorority sister, who is in a window 4.00 m above. The keys are caught 1.50 s later by the sister's outstretched hand.

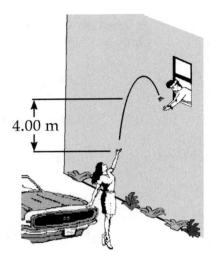

(a) With what initial velocity were the keys thrown?

(b) What was the velocity of the keys just before they were caught?

FIG. P2.31

Solution We model the keys as a particle under the constant free-fall acceleration. Taking the student's position to be $y_i = 0$, and given that $y = 4.00$ m at $t = 1.50$ s, we find (with $a = -9.80$ m/s^2).

(a) $y_f = y_i + v_{yi}t + \dfrac{1}{2}a_yt^2$ or $v_{yi} = \dfrac{y_f - y_i - \frac{1}{2}a_yt^2}{t}$

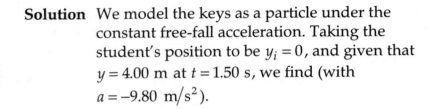

$$v_i = \frac{4.00 \text{ m} - \frac{1}{2}(-9.80 \text{ m/s}^2)(1.50 \text{ s})^2}{1.50 \text{ s}} = 10.0 \text{ m/s} \qquad \Diamond$$

(b) The velocity at any time $t > 0$ is given by $v_{yf} = v_{yi} + a_yt$. Therefore, at $t = 1.50$ s,

$$v_{yf} = 10.0 \text{ m/s} - \left(9.80 \text{ m/s}^2\right)(1.50 \text{ s}) = -4.68 \text{ m/s}. \qquad \Diamond$$

The negative sign means that the keys are moving **downward** just before they are caught.

P2.33 A daring ranch hand sitting on a tree limb wishes to drop vertically onto a horse galloping under the tree. The constant speed of the horse is 10.0 m/s, and the distance from the limb to the level of the saddle is 3.00 m.

(a) What must be the horizontal distance between the saddle and limb when the ranch hand makes his move?

(b) How long is he in the air?

continued on next page

Solution We do part (b) first.

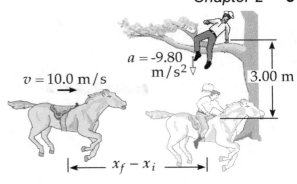

FIG. P2.33

(b)   Consider the vertical motion of the man after leaving the limb (with $v_i = 0$ at $y_i = 3.00$ m) until reaching the saddle (at $y_f = 0$). Modeling the man as a particle under constant acceleration, we find his time of fall from

$$y_f = y_i + v_{yi}t + \frac{1}{2}a_y t^2.$$

When $v_i = 0$,

$$t = \sqrt{\frac{2(y_f - y_i)}{a_y}} = \sqrt{\frac{2(0 - 3.00 \text{ m})}{-9.80 \text{ m/s}^2}} = 0.782 \text{ s}. \qquad \lozenge$$

(a)   During this time interval, the horse is modeled as a particle under constant velocity in the horizontal direction.

$$v_{xi} = v_{xf} = 10.0 \text{ m/s so } x_f - x_i = v_{xi}t = (10.0 \text{ m/s})(0.782 \text{ s}) = 7.82 \text{ m}$$

and the ranch hand must let go when the horse is 7.82 m from the tree. $\lozenge$

P2.39   For many years, Colonel John P. Stapp, USAF, held the world's land-speed record. He participated in studying whether a jet pilot could survive emergency ejection. On March 19, 1954, he rode a rocket-propelled sled that moved down a track at 632 mi/h. He and the sled were safely brought to rest in 1.40 s.

(a)   Determine the negative acceleration he experienced.

(b)   Determine the distance he traveled during this negative acceleration, assumed to be constant.

Solution **Conceptualize.** We estimate the acceleration as between $-10g$ and $-100g$: that is, between -100 m/s^2 and $-1\,000 \text{ m/s}^2$. We have already chosen the straight track as the x axis and the direction of travel as positive. We expect the stopping distance to be on the order of 100 m.

Categorize. We assume the acceleration is constant. We choose the initial and final points 1.40 s apart, bracketing the slowing-down process. Then we have a straightforward problem about a problem under constant acceleration.

continued on next page

Analyze. $v_{xi} = 632 \text{ mi/h} = 632 \text{ mi/h}\left(\dfrac{1\,609 \text{ m}}{1 \text{ mi}}\right)\left(\dfrac{1 \text{ h}}{3\,600 \text{ s}}\right) = 282 \text{ m/s}$

(a)   Taking $v_{xf} = v_{xi} + a_x t$ with $v_{xf} = 0$,

$$a_x = \frac{v_{xf} - v_{xi}}{t} = \frac{0 - 282 \text{ m/s}}{1.40 \text{ s}} = -202 \text{ m/s}^2 . \qquad \diamondsuit$$

This is approximately 20g.

(b) $x_f - x_i = \dfrac{1}{2}(v_{xi} + v_{xf})t = \dfrac{1}{2}(282 \text{ m/s} + 0)(1.40 \text{ s}) = 198 \text{ m}$ $\qquad \diamondsuit$

Finalize. While $x_f - x_i$, v_{xi}, and t are all positive, a_x is negative as expected. Our answers for a_x and for the distance agree with an order-of-magnitude estimates.

P2.49   Setting a new world record in a 100-m race, Maggie and Judy cross the finish line in a dead heat, both taking 10.2 s. Accelerating uniformly, Maggie took 2.00 s and Judy 3.00 s to attain maximum speed, which they maintained for the rest of the race.

(a)   What was the acceleration of each sprinter?

(b)   What were their respective maximum speeds?

(c)   Which sprinter was ahead at the 6.00-s mark, and by how much?

Solution (a)   Maggie moves with constant positive acceleration a_M for 2.00 s, then with constant speed (zero acceleration) for 8.20 s, covering a distance of $x_{M1} + x_{M2} = 100 \text{ m}$. The two component distances are

$$x_{M1} = \frac{1}{2}a_M(2.00 \text{ s})^2 \text{ and } x_{M2} = v_M(8.20 \text{ s}),$$

where v_M is her maximum speed, $v_M = 0 + a_M(2.00 \text{ s})$. By substitution

$$\frac{1}{2}a_M(2.00 \text{ s})^2 + a_M(2.00 \text{ s})(8.20 \text{ s}) = 100 \text{ m}$$

$$a_M = 5.43 \text{ m/s}^2 . \qquad \diamondsuit$$

continued on next page

Similarly, for Judy, $x_{J1} + x_{J2} = 100$ m with $x_{J1} = \frac{1}{2}a_J(3.00 \text{ s})^2$;

$x_{J2} = v_J(7.20 \text{ s})$; $v_J = a_J(3.00 \text{ s})$

$$\frac{1}{2}a_J(3.00 \text{ s})^2 + a_J(3.00 \text{ s})(7.20 \text{ s}) = 100 \text{ m}$$

$$a_J = 3.83 \text{ m/s}^2 \qquad \diamondsuit$$

(b)   Their speeds after accelerating are

$$v_M = a_M(2.00 \text{ s}) = (5.43 \text{ m/s}^2)(2.00 \text{ s}) = 10.9 \text{ m/s} \qquad \diamondsuit$$

and

$$v_J = a_J(3.00 \text{ s}) = (3.83 \text{ m/s}^2)(3.00 \text{ s}) = 11.5 \text{ m/s}. \qquad \diamondsuit$$

(c)   In the first 6.00 s, Maggie covers a distance

$$\frac{1}{2}a_M(2.00 \text{ s})^2 + v_M(4.00 \text{ s}) = \frac{1}{2}(5.43 \text{ m/s}^2)(2.00 \text{ s})^2 + (10.9 \text{ m/s})(4.00 \text{ s})$$
$$= 54.3 \text{ m}$$

and Judy has run a distance

$$\frac{1}{2}a_J(3.00 \text{ s})^2 + v_J(3.00 \text{ s}) = \frac{1}{2}(3.83 \text{ m/s}^2)(3.00 \text{ s})^2 + (11.5 \text{ m/s})(3.00 \text{ s}) = 51.7 \text{ m}$$

So Maggie is ahead by $54.3 \text{ m} - 51.7 \text{ m} = 2.62 \text{ m}$. $\qquad \diamondsuit$

P2.51   An inquisitive physics student and mountain climber climbs a 50.0-m cliff that overhangs a calm pool of water. He throws two stones vertically downward, 1.00 s apart, and observes that they cause a single splash. The first stone has an initial speed of 2.00 m/s.

(a)   How long after release of the first stone do the two stones hit the water?

(b)   What initial velocity must the second stone have if the two stones are to list simultaneously?

(c)   What is the speed of each stone at the instant the two hit the water?

continued on next page

Solution Set $y_i = 0$ at the top of the cliff, and find the time required for the first stone to reach the water using the particle under constant acceleration model:

$$y_f = y_i + v_{yi}t + \frac{1}{2}a_y t^2 \text{ or in quadratic form,}$$

$$-\frac{1}{2}a_y t^2 - v_{yi}t + y_f - y_i = 0.$$

(a) If we take the direction downward to be negative,

$$y_f = -50.0 \text{ m, } v_{yi} = -2.00 \text{ m/s, and } a_y = -9.80 \text{ m/s}^2.$$

Substituting these values into the equation, we find

$$\left(4.90 \text{ m/s}^2\right)t^2 + (2.00 \text{ m/s})t - 50.0 \text{ m} = 0.$$

Using the quadratic formula, and noting that only the positive root describes this physical situation,

$$t = \frac{-2.00 \text{ m/s} \pm \sqrt{(2.00 \text{ m/s})^2 - 4\left(4.90 \text{ m/s}^2\right)(-50.0 \text{ m})}}{2\left(4.90 \text{ m/s}^2\right)} = 3.00 \text{ s.} \qquad \lozenge$$

(b) For the second stone, the time interval of travel is $t = 3.00 \text{ s} - 1.00 \text{ s} = 2.00 \text{ s}$.
Since $y_f = y_i + v_{yi}t + \frac{1}{2}a_y t^2$,

$$v_{yi} = \frac{(y_f - y_i) - \frac{1}{2}a_y t^2}{t} = \frac{-50.0 \text{ m} - \left(\frac{1}{2}\right)\left(-9.80 \text{ m/s}^2\right)(2.00 \text{ s})^2}{2.00 \text{ s}} = -15.3 \text{ m/s.} \qquad \lozenge$$

The negative value indicates the downward direction of the initial velocity of the second stone.

continued on next page

(c) For the first stone, $v_{1f} = v_{1i} + a_1 t_1 = -2.00 \text{ m/s} + (-9.80 \text{ m/s}^2)(3.00 \text{ s})$

$v_{1f} = -31.4 \text{ m/s}.$ ◊

For the second stone, $v_{2f} = v_{2i} + a_2 t_2 = -15.3 \text{ m/s} + (-9.80 \text{ m/s}^2)(2.00 \text{ s})$

$v_{2f} = -34.8 \text{ m/s}.$ ◊

P2.57 Two objects, A and B, are connected by a rigid rod that has a length L. The objects slide along perpendicular guide rails, as shown in Figure P2.57. If A slides to the left with a constant speed v, find the speed of B when $\alpha = 60.0°$.

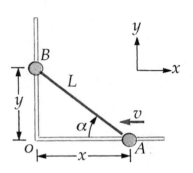

FIG. P2.57

Solution We translate from a pictorial representation through a geometric model to a mathematical representation by observing that the distances x and y are always related by $x^2 + y^2 = L^2$. Differentiating this equation with respect to time, we have

$$2x \frac{dx}{dt} + 2y \frac{dy}{dt} = 0.$$

Now the unknown velocity of B is $\qquad \dfrac{dy}{dt} = v_B$

and $\qquad \dfrac{dx}{dt} = -v.$

So the differentiated equation becomes $\qquad \dfrac{dy}{dt} = -\dfrac{x}{y}\left(\dfrac{dx}{dt}\right) = \left(-\dfrac{x}{y}\right)(-v) = v_B.$

But $\qquad \dfrac{y}{x} = \tan \alpha$

so $\qquad v_B = \left(\dfrac{1}{\tan \alpha}\right) v.$

When $\alpha = 60.0°$, $\qquad v_B = \dfrac{v}{\tan 60.0°} = \dfrac{v}{\sqrt{3}} = 0.577v.$ ◊

continued on next page

The form $v_B = \dfrac{v}{\tan\alpha}$ shows that the speed of slider B is directly proportional to the speed of A, but generally different. This is analogous to the way that different points on a rotating turntable have different linear speeds, all proportional to the turntable's rate of rotation. The inverse proportionally to $\tan\theta$ indicates that B moves much faster than A when A starts its motion at distance L from point O. Then B gradually slows down (with nonconstant acceleration) to come to rest as A approaches O.

Motion in Two Dimensions

Chapter 3

NOTES FROM SELECTED CHAPTER SECTIONS

Section 3.1   The Position, Velocity, and Acceleration Vectors

The **position vector** designates the location of a point relative to the origin of a coordinate system. The **displacement vector** is the difference between final and initial position vectors. The **velocity vector** is the rate at which the displacement changes with time and the **acceleration vector** is the rate at which the velocity changes with time.

It is important to recognize that a particle experiences an acceleration when:

- The magnitude of the velocity (speed) changes while the direction remains constant (e.g., a sphere rolling down an inclined plane).

- The velocity's magnitude remains constant while the velocity's direction changes (e.g., a particle moving at constant speed around a circle of constant radius).

- Both the magnitude and direction of the velocity change (e.g., a mass vibrating up and down at the end of a spring).

Section 3.3   Projectile Motion

Projectile motion is a common example of **motion in two dimensions** under constant acceleration. Provided air resistance is negligible, the characteristics of projectile motion can be summarized as follows:

- The horizontal component of velocity, v_x, remains constant since the horizontal component of acceleration is zero.

- The vertical component of acceleration is downward and equal to the acceleration due to gravity, g.

- The vertical component of velocity, v_y, and the displacement in the y direction change in time in a manner identical to that of a freely-falling body.

- Projectile motion can be described as a superposition, or vector addition, of the two motions in the x and y directions.

The **trajectory of a projectile is a parabola** as shown in Figure 3.1. We choose the motion to be in the x-y plane. The initial velocity of the projectile to have a magnitude of v_i, and is directed at an angle θ_i above the x-axis. The parabolic path of travel is completely determined when the magnitude, v_i, and the direction, θ_i, of the initial velocity vector are given. *The actual motion of the projectile is a superposition of two motions: motion of a freely-falling body with constant acceleration in the vertical direction, $-g$, and motion in the horizontal direction with constant velocity, $v_x = v_i \cos\theta_i$.*

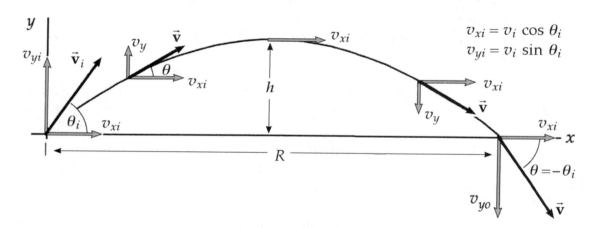

Figure 3.1 The **maximum height** h and **horizontal range** R are special coordinates defined here. They can be obtained from Equations 3.15 and 3.16.

Section 3.4   The Particle in Uniform Circular Motion

Centripetal acceleration is that acceleration experienced by a mass moving uniformly in a circular path of constant radius. *The direction of the centripetal acceleration is always toward the center of the circular path.*

Uniform circular motion is the motion of a particle moving in a circular path with constant linear speed. *The velocity vector is always tangent to the path of the moving body, and it is perpendicular to the acceleration vector directed toward the center of the circle.*

Section 3.5   Tangential and Radial Acceleration

Tangential acceleration of a particle moving in a circular path is due to a change in the speed (magnitude of the velocity vector) of the particle. *The direction of the tangential acceleration at any instant is tangent to the circular path (perpendicular to the radius).*

   If a particle moves in a circle such that the speed v is not constant, the total acceleration is the vector sum of the tangential and radial components. Figure 3.2 illustrates the case for a particle moving counterclockwise with increasing speed.

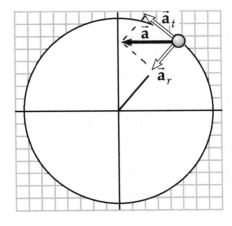

FIG. 3.2 Components of acceleration when $|\vec{v}|$ is not constant

EQUATIONS AND CONCEPTS

The **displacement** of a particle, $\Delta\vec{r} \equiv \vec{r}_f - \vec{r}_i$, is defined as the difference between the final and initial position vectors. *For the displacement illustrated in the figure, the magnitude of $\Delta\vec{r}$ is less than the actual path length from the initial to the final position.*

$$\Delta\vec{r} \equiv \vec{r}_f - \vec{r}_i \tag{3.1}$$

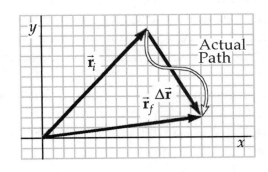

FIG. 3.3

The **average velocity** of a particle that undergoes a displacement $\Delta\vec{r}$ in a time interval Δt equals the ratio $\dfrac{\Delta\vec{r}}{\Delta t}$. *The average velocity depends on the displacement vector and not on the path traveled.*

$$\vec{v}_{\text{avg}} \equiv \frac{\Delta\vec{r}}{\Delta t} \tag{3.2}$$

The **instantaneous velocity** of a particle equals the limit of the average velocity as $\Delta t \to 0$. *The magnitude of the instantaneous velocity is called the speed.*

$$\vec{\mathbf{v}} \equiv \lim_{\Delta t \to 0} \frac{\Delta \vec{\mathbf{r}}}{\Delta t} = \frac{d\vec{\mathbf{r}}}{dt} \qquad (3.3)$$

The **average acceleration** of a particle that undergoes a change in velocity $\Delta \vec{\mathbf{v}}$ in a time interval Δt equals the ratio $\dfrac{\Delta \vec{\mathbf{v}}}{\Delta t}$.

$$\vec{\mathbf{a}}_{\text{avg}} \equiv \frac{\vec{\mathbf{v}}_f - \vec{\mathbf{v}}_i}{t_f - t_i} = \frac{\Delta \vec{\mathbf{v}}}{\Delta t} \qquad (3.4)$$

The **instantaneous acceleration** is defined as the limit of the average acceleration as $\Delta t \to 0$. *An object experiences an acceleration when the velocity vector undergoes a change in magnitude, direction, or both.*

$$\vec{\mathbf{a}} \equiv \lim_{\Delta t \to 0} \frac{\Delta \vec{\mathbf{v}}}{\Delta t} = \frac{d\vec{\mathbf{v}}}{dt} \qquad (3.5)$$

The **position vector** for a particle moving in the x-y plane can be written in terms of the coordinates, which change with time.

$$\vec{\mathbf{r}} = x\hat{\mathbf{i}} + y\hat{\mathbf{j}} \qquad (3.6)$$

Motion in two dimensions with constant acceleration is described by equations for velocity and position, which are vector versions of the one-dimensional kinematic equations.

$$\vec{\mathbf{v}}_f = \vec{\mathbf{v}}_i + \vec{\mathbf{a}}t \qquad (3.8)$$

$$\vec{\mathbf{r}}_f = \vec{\mathbf{r}}_i + \vec{\mathbf{v}}_i t + \frac{1}{2}\vec{\mathbf{a}}t^2 \qquad (3.9)$$

For projectile motion: Equations 3.10 and 3.11 give the x and y components of velocity and Equations 3.12 and 3.13 give the x and y position coordinates as functions of time assuming the projectile starts at the origin at $t = 0$. *A projectile moves along the horizontal with constant velocity and along the vertical with constant acceleration.*

$$v_{xf} = v_{xi} = v_i \cos\theta_i = \text{constant} \qquad (3.10)$$

$$v_{yf} = v_{yi} - gt = v_i \sin\theta_i - gt \qquad (3.11)$$

$$x_f = (v_i \cos\theta_i)t \qquad (3.12)$$

$$y_f = (v_i \sin\theta_i)t - \frac{1}{2}gt^2 \qquad (3.13)$$

The **trajectory (path) of a projectile** is a parabola. The path of the projectile is completely determined if the values of v_i and θ_i are known. Equation 3.14 is valid for angles in the range $0 < \theta_i < \dfrac{\pi}{2}$.

$$y_f = (\tan\theta_i)x_f - \left(\frac{g}{2v_i{}^2 \cos^2\theta_i}\right)x_f{}^2 \qquad (3.14)$$

The **maximum height** of a projectile can be written in terms of v_i and θ_i.

$$h = \frac{v_i^2 \sin^2 \theta_i}{2g} \tag{3.15}$$

The **horizontal range** of a projectile can also be stated in terms of v_i and θ_i.

$$R = \frac{v_i^2 \sin 2\theta_i}{g} \tag{3.16}$$

The **centripetal acceleration** vector for a particle in *uniform circular motion* is directed toward the center of the circular path.

$$a_c = \frac{v^2}{r} \tag{3.17}$$

The **period** (T) of a particle moving with constant speed in a circle of radius r is defined as the time required to complete one revolution.

$$T = \frac{2\pi r}{v} \tag{3.18}$$

The **total acceleration** of a particle moving on a curved path is the vector sum of the radial and the tangential components of acceleration (see Figure 3.2). The radial component, a_r is directed toward the center of curvature and arises from the change in direction of the velocity vector. The tangential component, a_t, is perpendicular to the radius and gives rise to the change in speed of the particle.

$$\vec{a} = \vec{a}_r + \vec{a}_t \tag{3.19}$$

$$a_r = -a_c = -\frac{v^2}{r}$$

$$a_t = \frac{d|\vec{v}|}{dt} \tag{3.20}$$

The **relative velocity, $\vec{v}_{PO}$** is the velocity of a particle as measured by a moving observer (moving at constant velocity with respect to another observer). Consider an observer O' moving with velocity $\vec{v}_{O'O}$ with respect to observer O, their measurements of the velocity of a particle located at point P are related according to Equations 3.21 and 3.22 (for one-dimensional motion).

$$\vec{v}_{PO} = \vec{v}_{PO'} + \vec{v}_{O'O} \tag{3.21}$$

$$v_{PO} = v_{PO'} + v_{O'O} \tag{3.22}$$

In Equations 3.21 and 3.22: First subscript describes what is being observed and the second describes who is doing the observing (e.g. $v_{O'O}$ is the velocity of O' as measured by O).

SUGGESTIONS, SKILLS, AND STRATEGIES

- You should be familiar with the mathematical expression for a parabola. In particular, the equation that describes the trajectory of a projectile moving under the influence of gravity is given by

$$y = Ax - Bx^2$$

where $A = \tan\theta_i$ and

$$B = \frac{g}{2v_i^2 \cos^2 \theta_i}.$$

Note that this expression for y assumes that the particle leaves the origin at $t = 0$, with a velocity $\vec{v}_i$ which makes an angle θ_i with the horizontal. A sketch of y versus x for this situation is shown in Figure 3.4.

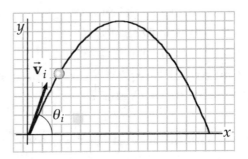

FIG. 3.4

- If you are given v_i and θ_i, you should be able to make a point-by-point plot of the trajectory using the expressions for $x(t)$ and $y(t)$. Furthermore, you should know how to calculate the velocity component v_y at any time t. (Note that the component $v_x = v_{xi} = v_i \cos\theta_i = \text{constant}$, since $a_x = 0$.)

- Assuming that you have values for x and y at any time $t > 0$, you should be able to write an expression for the position vector $\vec{r}$ at that time using the relation $\vec{r} = x\hat{i} + y\hat{j}$. From this you can find the magnitude of the displacement r, where $r = \sqrt{x^2 + y^2}$. Likewise, if v_x and v_y are known at any time $t > 0$, you can express the velocity vector $\vec{v}$ in the formula $\vec{v} = v_x\hat{i} + v_y\hat{j}$. From this, you can find the speed at any time, since $v = \sqrt{v_x^2 + v_y^2}$.

Problem-Solving Strategy: Projectile Motion

You should use the following approach to solving projectile motion problems:

- Select a coordinate system.

- Resolve the initial velocity vector into x and y components.

- Treat the horizontal motion and the vertical motion independently.

- Follow the techniques for solving problems with constant velocity to analyze the horizontal motion of the projectile.

- Follow the techniques for solving problems with constant acceleration to analyze the vertical motion of the projectile.

REVIEW CHECKLIST

✓ Describe the displacement, velocity, and acceleration of a particle moving in the x-y plane.

✓ Recognize that two-dimensional motion in the x-y plane with constant acceleration is equivalent to two independent motions along the x and y directions with constant acceleration components a_x and a_y.

✓ Recognize the fact that if the initial speed v_i and initial angle θ_i of a projectile are known at a given point at $t = 0$, the velocity components and coordinates can be found at any later time t. Furthermore, one can also calculate the horizontal range R and maximum height h if v_i and θ_i are known.

✓ Understand the nature of the acceleration of a particle moving in a circle with constant speed. In this situation, note that although $|\vec{v}|$ = constant, the direction of $\vec{v}$ varies in time, the result of which is the radial, or centripetal acceleration.

✓ Describe the components of acceleration for a particle moving on a curved path, where both the magnitude and direction of $\vec{v}$ are changing with time. *In this case, the particle has a tangential component of acceleration and a radial component of acceleration.*

✓ Realize that the outcome of a measurement of the motion of a particle (its position, velocity, and acceleration) depends on the frame of reference of the observer.

ANSWERS TO SELECTED QUESTIONS

Q3.1 If you know the position vectors of a particle at two points along its path and also know the time interval it took to move from one point to the other, can you determine the particle's instantaneous velocity? Its average velocity? Explain.

Answer Its instantaneous velocity cannot be determined at any point from this information. However, the average velocity over the time interval can be determined from its definition and the given information.

Q3.5 A spacecraft drifts through space at a constant velocity. Suddenly a gas leak in the side of the spacecraft gives it a constant acceleration in a direction perpendicular to the initial velocity. The orientation of the spacecraft does not change, so the acceleration remains perpendicular to the original direction of the velocity. What is the shape of the path followed by the spacecraft in this situation?

Answer The spacecraft will follow a parabolic path. This is equivalent to a projectile thrown off a cliff with a horizontal velocity. For the projectile, gravity provides an acceleration that is always perpendicular to the initial velocity, resulting in a parabolic path. For the spacecraft, the initial velocity plays the role of the horizontal velocity of the projectile. The leaking gas provides an acceleration that plays the role of gravity for the projectile. If the orientation of the spacecraft were to change in response to the gas leak (which is by far the more likely result), then the acceleration would change direction and the motion could become quite complicated.

Q3.7 A projectile is launched at some angle to the horizontal with some initial speed v_i, and air resistance is negligible. Is the projectile a freely falling body? What is its acceleration in the vertical direction? What is its acceleration in the horizontal direction?

Answer Yes. The acceleration is that of a freely falling body, because nothing counteracts the force of gravity. The vertical acceleration will be the local gravitational acceleration, g; the horizontal acceleration will be zero.

Q3.11 Explain whether or not the following particles have an acceleration:

    (a) A particle moving in a straight line with constant speed.

    (b) A particle moving around a curve with constant speed.

Answer (a) The acceleration is zero, since the magnitude and direction of $\vec{v}$ remain constant.

    (b) The particle has an acceleration since the direction of $\vec{v}$ changes.

SOLUTIONS TO SELECTED PROBLEMS

P3.1   A motorist drives south at 20.0 m/s for 3.00 min, then turns west and travels at 25.0 m/s for 2.00 min, and finally travels northwest at 30.0 m/s for 1.00 min. For this 6.00-min trip,

(a)   Find the total vector displacement.

(b)   Find the average speed.

(c)   Find the average velocity. Let the positive *x*-axis point east.

Solution (a)   For each segment of the motion we model the car as a particle under constant velocity. Her displacements are

$$\Delta \vec{r} = (20.0 \text{ m/s})(180 \text{ s}) \text{ south} + (25.0 \text{ m/s})(120 \text{ s}) \text{ west}$$
$$+ (30.0 \text{ m/s})(60.0 \text{ s}) \text{ northwest}$$

Choosing $\hat{\mathbf{i}}$ = east and $\hat{\mathbf{j}}$ = north, we have

$$\Delta \vec{r} = (3.60 \text{ km})(-\hat{\mathbf{j}}) + (3.00 \text{ km})(-\hat{\mathbf{i}}) + (1.80 \text{ km cos } 45.0°)(-\hat{\mathbf{i}})$$
$$+ (1.80 \text{ km sin } 45.0°)(\hat{\mathbf{j}})$$
$$\Delta \vec{r} = (3.00 + 1.27) \text{ km } (-\hat{\mathbf{i}}) + (1.27 - 3.60) \text{ km } \hat{\mathbf{j}} = (-4.27\hat{\mathbf{i}} - 2.33\hat{\mathbf{j}}) \text{ km} \qquad \Diamond$$

The answer can also be written as

$$\Delta \vec{r} = \sqrt{(-4.27 \text{ km})^2 + (-2.33 \text{ km})^2} \text{ at } \tan^{-1}\left(\frac{2.33}{4.27}\right) = 29° \text{ south of west,}$$

or $\Delta \vec{r} = 4.87$ km at 209° from east $\Diamond$

(b)   The total distance or path-length traveled is $(3.60 + 3.00 + 1.80) \text{ km} = 8.40 \text{ km}$ so

$$\text{average speed} = \frac{8.40 \text{ km}}{6.00 \text{ min}}\left(\frac{1.00 \text{ min}}{60.0 \text{ s}}\right)\left(\frac{1000 \text{ m}}{\text{km}}\right) = 23.3 \text{ m/s} \qquad \Diamond$$

(c) $\vec{\mathbf{v}}_{\text{av}} = \dfrac{\Delta \vec{r}}{t} = \dfrac{4.87 \text{ km}}{360 \text{ s}} = 13.5 \text{ m/s at } 209°$

or $\vec{\mathbf{v}}_{\text{av}} = \dfrac{\Delta \vec{r}}{t} = \dfrac{(-4.27 \text{ east} - 2.33 \text{ north}) \text{ km}}{360 \text{ s}} = (11.9 \text{ west} + 6.47 \text{ south}) \text{ m/s} \quad \Diamond$

P3.3   A fish swimming in a horizontal plane has velocity $\vec{v}_i = (4.00\hat{i} + 1.00\hat{j})$ m/s at a point in the ocean where the position relative to a certain rock is $\vec{r}_i = (10.0\hat{i} - 4.00\hat{j})$ m. After the fish swims with constant acceleration for 20.0 s, its velocity is $\vec{v} = (20.0\hat{i} - 5.00\hat{j})$ m/s.

(a)   What are the components of the acceleration?

(b)   What is the direction of the acceleration with respect to unit vector $\hat{i}$?

(c)   If the fish maintains constant acceleration where is the fish at $t = 25.0$ s and in what direction is it moving?

Solution Model the fish as a particle under constant acceleration.

At $t = 0$, $\vec{v}_i = (4.00\hat{i} + 1.00\hat{j})$ m/s and $\vec{r}_i = (10.0\hat{i} - 4.00\hat{j})$ m

$\vec{v}_f = (20.0\hat{i} - 5.00\hat{j})$ m/s

(a) $a_x = \dfrac{\Delta v_x}{\Delta t} = \dfrac{20.0 \text{ m/s} - 4.00 \text{ m/s}}{20.0 \text{ s}} = 0.800 \text{ m/s}^2$ ◊

$a_y = \dfrac{\Delta v_y}{\Delta t} = \dfrac{-5.00 \text{ m/s} - 1.00 \text{ m/s}}{20.0 \text{ s}} = -0.300 \text{ m/s}^2$ ◊

(b) $\theta = \tan^{-1}\left(\dfrac{a_y}{a_x}\right) = \tan^{-1}\left(\dfrac{-0.300 \text{ m/s}^2}{0.800 \text{ m/s}^2}\right) = -20.6°$ or $339°$ from the $+x$ axis   ◊

(c)   At $t = 25.0$ s, its coordinates are $x_f = x_i + v_{xi}t + \dfrac{1}{2}a_x t^2$ and

$y_f = y_i + v_{yi}t + \dfrac{1}{2}a_y t^2$

$x_f = x_i + v_{xi}t + \dfrac{1}{2}a_x t^2 = 10.0 \text{ m} + (4.00 \text{ m/s})(25.0 \text{ s}) + \dfrac{1}{2}(0.800 \text{ m/s}^2)(25.0 \text{ s})^2$

$x_f = 360 \text{ m}$ ◊

$y_f = y_i + v_{yi}t + \dfrac{1}{2}a_y t^2 = -4.00 \text{ m} + (1.00 \text{ m/s})(25.0 \text{ s}) + \dfrac{(-0.300 \text{ m/s}^2)(25.0 \text{ s})^2}{2}$

$y_f = -72.8 \text{ m}$ ◊

continued on next page

$$v_{xf} = v_{xi} + a_x t = (4.00 \text{ m/s}) + (0.800 \text{ m/s}^2)(25.0 \text{ s}) = 24.0 \text{ m/s}$$

$$v_{yf} = v_{yi} + a_y t = (1.00 \text{ m/s}) - (0.300 \text{ m/s}^2)(25.0 \text{ s}) = -6.50 \text{ m/s}$$

Therefore, $\vec{r} = (360\hat{i} - 72.8\hat{j}) \text{ m}$

$$\theta = \tan^{-1}\left(\frac{v_y}{v_x}\right) = \tan^{-1}\left(\frac{-6.50 \text{ m/s}}{24.0 \text{ m/s}}\right) = -15.2° = 345° \text{ from the } +x \text{ axis.} \quad \lozenge$$

P3.7   In a local bar, a customer slides an empty beer mug down the counter for a refill. The bartender is just deciding to go home and rethink his life does not see the mug, which slides off the counter and strikes the floor 1.40 m from the base of the counter. If the height of the counter is 0.860 m,

(a)   With what velocity did the mug leave the counter?

(b)   What was the direction of the mug's velocity just before it hit the floor?

Solution Conceptualize: Based on our everyday experiences and the description of the problem, a reasonable speed of the mug would be a few m/s and it will hit the floor at some angle between 0° and 90°, probably about 45°.

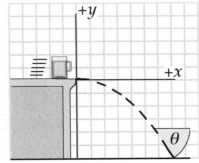

FIG. P3.7

Categorize: We are looking for two different velocities, but we are only given two distances. Our approach will be to separate the vertical and horizontal motions. By using the height that the mug falls, we can find the time of the fall. Once we know the time, we can find the horizontal and vertical components of the velocity. For convenience, we will set the origins to be the point where the mug leaves the counter.

Analyze: Vertical motion: $y = -0.860 \text{ m}$ $v_{yi} = 0$ $v_y = ?$ $a_y = -9.80 \text{ m/s}^2$

Horizontal motion: $x = 1.40 \text{ m}$ $v_x = ? = \text{constant}$ $a_x = 0$

continued on next page

(a) To find the time of fall, we use the free fall equation: $y = v_{yi}t + \frac{1}{2}a_y t^2$.

Solving, $-0.860 \text{ m} = 0 + \frac{1}{2}(-9.80 \text{ m/s}^2)t^2$ and $t = 0.419$ s. Then,

$$v_x = \frac{x}{t} = \frac{1.40 \text{ m}}{0.419 \text{ s}} = 3.34 \text{ m/s}. \qquad \Diamond$$

(b) The mug hits the floor with a vertical velocity of $v_{yf} = v_{yi} + a_y t$ and an

impact angle below the horizontal of $\theta = \tan^{-1}\left(\frac{|v_y|}{v_x}\right)$. Solving for v_y,

$$v_y = 0 - \left(9.80 \text{ m/s}^2\right)(0.419 \text{ s}) = -4.11 \text{ m/s}.$$

Thus, $\theta = \tan^{-1}\left(\frac{4.11 \text{ m/s}}{3.34 \text{ m/s}}\right) = 50.9°. \qquad \Diamond$

Finalize: This was a multi-step problem that required several physics equations to solve; yet our answers do agree with our initial expectations. Since the problem did not ask for the time, we could have eliminated this variable by substitution, but then we would have had to substitute the algebraic expression $t = \sqrt{\frac{2y}{g}}$ into two other equations. So in this case it was easier to find a numerical value for the time as an intermediate step. Sometimes the most efficient method is not realized until each alternative solution is attempted.

P3.15 A placekicker must kick a football from a point 36.0 m (about 40 yards) from the goal, and half the crowd hopes the ball will clear the crossbar, which is 3.05 m high. When kicked, the ball leaves the ground with a speed of 20.0 m/s at an angle of 53.0° to the horizontal.

(a) By how much does the ball clear or fall short of clearing the crossbar?

(b) Does the ball approach the crossbar while still rising or while falling?

Solution Model the football as a projectile, moving with constant horizontal velocity and with constant vertical acceleration.

continued on next page

(a)   To find the actual height of the football when it reaches the goal line, we can use the trajectory equation:

$$y_f = x_f \tan\theta_i - \frac{gx_f^2}{2v_i^2 \cos^2\theta_i}$$

because we are given $x_f = 36.0$ m, $v_i = 20.0$ m/s, and $\theta_i = 53.0°$. So,

$$y_f = (36.0 \text{ m})(\tan 53.0°) - \frac{(9.80 \text{ m/s}^2)(36.0 \text{ m})^2}{(2)(20.0 \text{ m/s})^2 \cos^2 53.0°}$$

$$y_f = 47.774 - 43.834 = 3.939 \text{ m}$$

The ball clears the bar by $(3.939 - 3.050)$ m $= 0.889$ m.   ◊

(b)   The time the ball takes to reach the maximum height $(v_y = 0)$ is

$$t_1 = \frac{(v_i \sin\theta_i) - v_y}{g} = \frac{(20.0 \text{ m/s})(\sin 53.0°) - 0}{9.80 \text{ m/s}^2} = 1.63 \text{ s}$$

The time to travel 36.0 m horizontally is

$$t_2 = \frac{36.0 \text{ m}}{(20.0 \text{ m/s})(\cos 53.0°)} = 2.99 \text{ s}.$$

Since $t_2 > t_1$, the ball clears the goal on its way down.   ◊

P3.23   The athlete shown in Figure P3.23 of the textbook rotates a 1.00-kg discus along a circular path of radius 1.06 m. The maximum speed of the discus is 20.0 m/s. Determine the magnitude of the maximum radial acceleration of the discus.

Solution The maximum radial acceleration occurs when maximum tangential speed is attained. Model the discus here as a particle in uniform circular motion. Here,

$$a_c = \frac{v^2}{r} = \frac{(20.0 \text{ m/s})^2}{(1.06 \text{ m})} = 377 \text{ m/s}^2$$ ◊

P3.29 A train slows down as it rounds a sharp horizontal turn, slowing from 90.0 km/h to 50.0 km/h in the 15.0 s that it takes to round the bend. The radius of the curve is 150 m. Compute the acceleration at the moment the train speed reaches 50.0 km/h. Assume that it continues to slow down at this time at the same rate.

Solution **Conceptualize:** If the train is taking this turn at a safe speed, then its acceleration should be significantly less than g, perhaps a few m/s^2 (otherwise it might jump the tracks!), and it should be directed toward the center of the curve and backwards since the train is slowing.

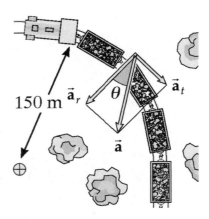

Categorize: Since the train is changing both its speed and direction, the acceleration vector will be the vector sum of the tangential and radial acceleration components. The tangential acceleration can be found from the changing speed and elapsed time, while the radial acceleration can be found from the radius of curvature and the train's speed.

FIG. P3.29

Analyze: First, let's convert the speed units from km/h to m/s:

$$v_i = 90.0 \ km/h = (90.0 \ km/h)(10^3 \ m/km)\left(\frac{1 \ h}{3\ 600 \ s}\right) = 25.0 \ m/s$$

$$v_f = 50.0 \ km/h = (50.0 \ km/h)(10^3 \ m/km)\left(\frac{1 \ h}{3\ 600 \ s}\right) = 13.9 \ m/s$$

The tangential acceleration and radial acceleration are:

$$a_t = \frac{\Delta v}{\Delta t} = \frac{13.9 \ m/s - 25.0 \ m/s}{15.0 \ s} = -0.741 \ m/s^2 \ \text{(backward)}$$

$$|a_r| = \frac{v^2}{r} = \frac{(13.9 \ m/s)^2}{150 \ m} = 1.29 \ m/s^2 \ \text{(inward)}$$

$$a = \sqrt{a_r^2 + a_t^2} = \sqrt{(1.29 \ m/s^2)^2 + (-0.741 \ m/s^2)^2} = 1.48 \ m/s^2 \qquad \lozenge$$

$$\theta = \tan^{-1}\left(\frac{a_t}{a_r}\right) = \tan^{-1}\left(\frac{0.741 \ m/s^2}{1.29 \ m/s^2}\right) = 29.9° \qquad \lozenge$$

Finalize: The acceleration is clearly less than g, and it appears that most of the acceleration comes from the radial component, so it makes sense that the acceleration vector should point mostly inward toward the center of the curve and slightly backward due to the negative tangential acceleration.

P3.31   Figure P3.31 represents the total acceleration of a particle moving clockwise in a circle of radius 2.50 m at a certain instant of time. At this instant,

   (a)   find the radial acceleration.

   (b)   find the speed of the particle.

   (c)   find its tangential acceleration.

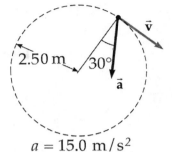

$a = 15.0 \ \text{m/s}^2$

FIG. P3.31

Solution (a)   The acceleration has an inward radial component:

$$a_c = a\cos 30.0° = \left(15.0 \ \text{m/s}^2\right)\cos 30.0° = 13.0 \ \text{m/s}^2$$ ◊

   (b)   The speed at the instant shown can be found by using

$$a_c = \frac{v^2}{r} \ \text{or} \ v = \sqrt{a_c r} = \sqrt{(13.0 \ \text{m/s}^2)(2.50 \ \text{m})} = 5.70 \ \text{m/s}.$$ ◊

   (c)   The acceleration also has a tangential component:

$$a_t = a\sin 30.0° = (15.0 \ \text{m/s}^2)\sin 30.0° = 7.50 \ \text{m/s}^2$$ ◊

P3.33   A river has a steady speed of 0.500 m/s. A student swims upstream a distance of 1.00 km and swims back to the starting point. If the student can swim at a speed of 1.20 m/s in still water, how long does the trip take? Compare this answer with the time interval the trip would take if the water were still.

Solution **Conceptualize:** If we think about the time for a trip as a function of the stream's speed, we realize that if the stream is flowing at the same rate or faster than the student can swim, he will never reach the 1.00-km mark even after an infinite amount of time. Since the student can swim 1.20 km in 1 000 s, we should expect that the trip will definitely take longer than in still water, maybe about 2 000 s (~30 minutes).

Categorize: The total time in the river is the longer time upstream (against the current) plus the shorter time downstream (with the current). For each part, we will use the basic equation $t = \frac{d}{v}$, where v is the speed of the student relative to the shore.

continued on next page

Analyze:

$$t_{up} = \frac{d}{v_{student} - v_{stream}} = \frac{1\,000 \text{ m}}{1.20 \text{ m/s} - 0.500 \text{ m/s}} = 1\,430 \text{ s}$$

$$t_{dn} = \frac{d}{v_{student} + v_{stream}} = \frac{1\,000 \text{ m}}{1.20 \text{ m/s} + 0.500 \text{ m/s}} = 588 \text{ s}$$

Total time in river, $t_{river} = t_{up} + t_{dn} = 2.02 \times 10^3$ s.   ◊

In still water, $t_{still} = \frac{d}{v} = \frac{2\,000 \text{ m}}{1.20 \text{ m/s}} = 1.67 \times 10^3$ s or

$$t_{still} = 0.827 t_{river}.$$ ◊

Finalize: As we predicted, it does take the student longer to swim up and back in the moving stream than in still water (21% longer in this case), and the amount of time agrees with our estimate.

P3.37   A science student is riding on a flatcar of a train traveling along a straight horizontal track at a constant speed of 10.0 m/s. The student throws a ball into the air along a path that he judges to make an initial angle of 60.0° with the horizontal and to be in line with the track. The student's professor, who is standing on the ground nearby, observes the ball to rise vertically. How high does she see the ball rise?

Solution   Shown on the right, $\vec{v}_{be} = \vec{v}_{ce} + \vec{v}_{bc}$ with $\vec{v}_{bc}$ = the velocity of the ball relative to the car, $\vec{v}_{be}$ = the velocity of the ball relative to the Earth and $\vec{v}_{ce}$ = car velocity relative to the Earth = 10.0 m/s. From the figure, we have $v_{ce} = v_{bc} \cos 60.0°$.

So $v_{bc} = \dfrac{10.0 \text{ m/s}}{\cos 60.0°} = 20.0 \text{ m/s}.$

Again from the figure,

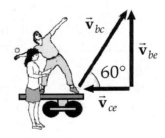

FIG. P3.37

$$v_{be} = v_{bc} \sin 60.0° + 0 = (20.0 \text{ m/s})(0.866) = 17.3 \text{ m/s}.$$

This is the initial velocity of the ball relative to the Earth. Now we can calculate the maximum height that the ball rises. From $v_{yf}^2 = v_{yi}^2 + 2ah$,

$$0 = (17.3 \text{ m/s})^2 + 2(-9.80 \text{ m/s}^2)h$$
$$h = 15.3 \text{ m}$$ ◊

P3.45   Barry Bonds hits a home run so that the baseball just clears the top row of bleachers, 21.0 m high, located 130 m from home plate. The ball is hit at an angle of 35.0° to the horizontal, and air resistance is negligible. Find

   (a)   the initial speed of the ball,

   (b)   the time interval that elapses before the ball reaches the top row, and

   (c)   the velocity components and the speed of the ball when it passes over the top row. (Assume the ball is hit at a height of 1.00 m above the ground.)

Solution   Let the initial speed of the ball be v_i, and the initial angle $\theta_i = 35.0°$. Set the starting point at the origin, $(x_i, y_i) = (0 \text{ m}, 0 \text{ m})$. When $x_f = 130$ m, $y_f = 20.0$ m.

$$x_f = x_i + v_{xi}t = v_i t \cos 35.0°$$

$$v_i t \cos 35.0° = 130 \text{ m}$$

and

$$v_i t = \frac{130 \text{ m}}{\cos 35.0°} = 158.7 \text{ m}$$

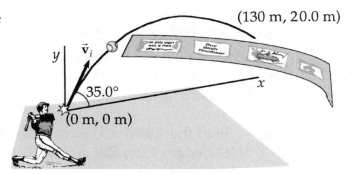

(130 m, 20.0 m)

$$v_{xi} = v_i \cos 35.0°$$
$$v_{yi} = v_i \sin 35.0°$$

FIG. P3.45

Next,

$$y_f = v_{yi}t - \frac{1}{2}gt^2 = (v_i \sin 35.0°)t - \frac{1}{2}(9.80 \text{ m/s}^2)t^2$$

Substituting for $v_i t$,

$$20.0 \text{ m} = (158.7 \text{ m})(\sin 35.0°) - \frac{1}{2}(9.80 \text{ m/s}^2)t^2$$

(b)   and $t = \sqrt{\dfrac{71.0 \text{ m}}{4.90 \text{ m/s}^2}} = 3.81 \text{ s}$ ◊

(a)   Therefore, $v_i = \dfrac{158.7 \text{ m}}{3.81 \text{ s}} = 41.7 \text{ m/s}$ ◊

continued on next page

(c) $\quad v_{yf} = v_{yi} - gt = v_i \sin 35.0° - gt = (23.9 \text{ m/s}) - (9.80 \text{ m/s}^2)t$

At $t = 3.81$ s, $v_y = -13.4$ m/s ◊

$$v_{xf} = v_{xi} = v_i \cos 35.0° = (41.7 \text{ m/s})\cos 35.0° = 34.1 \text{ m/s}$$ ◊

and $\qquad |\vec{v}| = \sqrt{v_x^2 + v_y^2} = \sqrt{34.1^2 + (-13.4)^2} = 36.7 \text{ m/s}$ ◊

P3.51 A car is parked on a steep incline overlooking the ocean, where the incline makes an angle of 37.0° below the horizontal. The negligent driver leaves the car in neutral, and the parking brakes are defective. Starting from rest at $t = 0$, the car rolls down the incline with a constant acceleration of 4.00 m/s^2, traveling 50.0 m to the edge of a vertical cliff. The cliff is 30.0 m above the ocean.

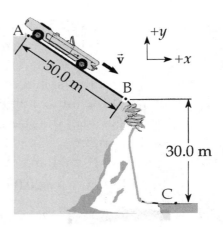

(a) Find the speed of the car when it reaches the edge of the cliff and the time interval it takes to get there.

FIG. P3.51

(b) Find the velocity of the car when it lands in the ocean.

(c) Find the total time interval during which the car is in motion.

(d) Find the position of the car when it lands in the ocean, relative to the base of the cliff.

Solution From point A to point B (along the incline), the car can be modeled as a particle under constant acceleration in one dimension, starting from rest ($v_i = 0$). Therefore, taking s to be the position along the incline,

(a) $\quad v_B^2 = v_i^2 + 2a(s_f - s_i) = 2a(s_B - s_A)$

$v_B = \sqrt{(2)(4.00 \text{ m/s}^2)(50.0 \text{ m})} = 20.0 \text{ m/s}$ ◊

We can find the elapsed time from $v_B = v_i + at$:

$$t_{AB} = \frac{v_B - v_i}{a} = \frac{20.0 \text{ m/s} - 0}{4.00 \text{ m/s}^2} = 5.00 \text{ s}$$ ◊

continued on next page

(b) After the car passes the top of the cliff, it becomes a projectile. At the edge of the cliff, the components of velocity v_B are:

$$v_{yB} = (-20.0 \text{ m/s}) \sin 37.0° = -12.0 \text{ m/s}$$
$$v_{xB} = (20.0 \text{ m/s}) \cos 37.0° = 16.0 \text{ m/s}$$

There is no further horizontal acceleration, so $v_{xC} = v_{xB} = 16.0$ m/s. However, the downward (negative) vertical velocity is affected by free fall:

$$v_{yC} = -\sqrt{2a_y(\Delta y) + v_{yB}{}^2} = -\sqrt{2(-9.80 \text{ m/s}^2)(-30.0 \text{ m}) + (-12.0 \text{ m/s})^2}$$
$$= -27.1 \text{ m/s}$$
$$\vec{v}_C = 16.0\hat{i} - 27.1\hat{j} \text{ m/s}$$

◊

(c) From point B to C, the time

$$t_{BC} = \frac{v_{yC} - v_{yB}}{a_y} = \frac{(-27.1 \text{ m/s}) - (-12.0 \text{ m/s})}{(-9.80 \text{ m/s}^2)} = 1.53 \text{ s}$$

The total elapsed time is $t_{AC} = t_{AB} + t_{BC} = 6.53$ s.

◊

(d) The horizontal distance covered is
$$\Delta x = v_{xB}t_{BC} = (16.0 \text{ m/s})(1.53 \text{ s}) = 24.5 \text{ m}.$$

◊

P3.57 A skier leaves the ramp of a ski jump with a velocity of 10.0 m/s, 15.0° above the horizontal, as in Figure P3.57. The slope is inclined at 50.0°, and air resistance is negligible.

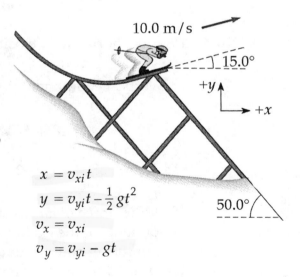

(a) Find the distance from the ramp to where the jumper lands.

$$x = v_{xi}t$$
$$y = v_{yi}t - \frac{1}{2}gt^2$$
$$v_x = v_{xi}$$
$$v_y = v_{yi} - gt$$

(b) Find the velocity components just before the landing. (How do you think the results might be affected if air resistance were included? Note that jumpers lean forward in the shape of an airfoil, with their hands at their sides, to increase their distance. Why does this work?)

FIG. P3.57

continued on next page

Solution Set point '0' where the skier takes off, and point '2' where the skier lands. Define the coordinate system as shown in the figure. $\vec{v}_i = 10.0$ m/s at 15.0°:

$$v_{xi} = v_i \cos\theta_i = (10.0 \text{ m/s})\cos 15.0° = 9.66 \text{ m/s}$$
$$v_{yi} = v_i \sin\theta_i = (10.0 \text{ m/s})\sin 15.0° = 2.59 \text{ m/s}$$

(a)   The skier travels horizontally $x_2 - x_0 = v_{xi}t = (9.66 \text{ m/s})t$ (1)

and vertically $y_2 - y_0 = v_{yi}t - \dfrac{1}{2}gt^2 = (2.59 \text{ m/s})t - (4.90 \text{ m/s}^2)t^2$ (2)

The skier hits the slope when $\dfrac{y_2 - y_0}{x_2 - x_0} = \tan(-50.0°) = -1.19$ (3)

Substituting (1) and (2) into (3), $\dfrac{(2.59 \text{ m/s})t - (4.90 \text{ m/s}^2)t^2}{(9.66 \text{ m/s})t} = -1.19$. Since

we ignore the solution $t = 0$, $4.90t - 14.1 = 0$ and $t = 2.88$ s. Solving Equation (1),

$$x_2 - x_0 = (9.66 \text{ m/s})t = 27.8 \text{ m}.$$

From the diagram $d = \dfrac{x_2 - x_0}{\cos 50.0°} = \dfrac{27.8 \text{ m}}{\cos 50.0°} = 43.2 \text{ m}.$ ◊

(b)   The final horizontal velocity is $v_{xf} = v_{xi} = 9.66 \text{ m/s}$.   ◊

The vertical component is found from

$$v_{yf} = v_{yi} - gt = 2.59 \text{ m/s} - (9.80 \text{ m/s}^2)t.$$

When $t = 2.88$ s, $v_{yf} = -25.6$ m/s.   ◊

The 'drag' force of air resistance would necessarily decrease both components of the ski jumper's impact velocity. On the other hand, the lift force of the air could extend her time of flight and increase the distance of her jump. If the jumper has the profile of an airplane wing, she can deflect downward the air through which she passes, to make the air deflect her upward.

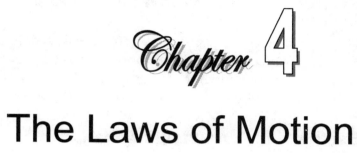

The Laws of Motion

Section 4.1   The Concept of Force

Equilibrium is the condition under which the net force (vector sum of all forces) acting on an object is zero. An object in equilibrium has a zero acceleration (velocity is constant or remains equals to zero).

Fundamental forces in nature are: (1) gravitational (attractive forces between objects due to their masses), (2) electromagnetic forces (between electric charges at rest or in motion), (3) strong nuclear forces (between subatomic particles), and (4) weak nuclear forces (accompanying the process of radioactive decay).

Classical physics is concerned with contact forces (which are the result of physical contact between two or more objects) and field forces (which act through empty space and do not involve physical contact).

Section 4.2   Newton's First Law

Section 4.3   Mass

Newton's first law is called the **law of inertia** and states that, when viewed from an inertial reference frame, an object at rest will remain at rest and an object in motion will remain in motion with a constant velocity unless acted on by a net external force. Inertial mass is a measure of an object's resistance to a change in motion, in response to an external force. Mass is an inherent property of an object, independent of the surroundings. *Mass and weight are different quantities; the weight of an object is equal to the magnitude of the gravitational force exerted on the object.*

Section 4.4 Newton's Second Law—The Particle Under a Net Force

Newton's second law, the **law of acceleration**, states that the acceleration of an object is directly proportional to the net force acting on it and inversely proportional to its mass. The direction of the acceleration is the direction of the net force. *The net force is also called the resultant force, total force, or sum of all forces.*

Section 4.6 Newton's Third Law

Newton's third law, the **law of action-reaction**, states that when two bodies interact, the force which body "1" exerts on body "2" (the **action force**) is equal in magnitude and opposite in direction to the force which body "2" exerts on body "1" (the **reaction force**). *A consequence of the third law is that forces occur in pairs; the action force and the reaction force act on different objects and therefore cannot cancel each other.*

Section 4.7 Applications of Newton's Laws

Construction of a **free-body diagram** is an important step in the application of Newton's laws of motion to the solution of problems involving bodies in equilibrium or accelerating under the action of external forces. The diagram should include an arrow labeled to identify each of the external forces acting on the body whose motion (or condition of equilibrium) is to be studied. *Forces which are the reactions to these external forces must not be included. When a system consists of more than one body or mass, you must construct a free-body diagram for each mass.*

EQUATIONS AND CONCEPTS

A **quantitative measurement of mass** (the term used to measure inertia) can be made by comparing the accelerations that a given force will produce on different bodies. If a given force acting on a body of mass m_1 produces an acceleration a_1 and the same force acting on a body of mass m_2 produces an acceleration a_2, the ratio of the two masses equals the inverse of the ratio of the two accelerations. *Mass is an inherent property of an object and is independent of the surroundings and the method of measurement.*

$$\frac{m_1}{m_2} = \frac{a_2}{a_1} \tag{4.1}$$

The **acceleration** of an object is proportional to the net force acting on it and inversely proportional to its mass. *This is a statement of Newton's second law.*

$$\sum \vec{F} = m\vec{a} \qquad (4.2)$$

Three **component equations** are the equivalent of the vector equation expressing Newton's second law. *The orientation of the coordinate system can often be chosen so that the object has a nonzero acceleration along only one direction.*

$$\Sigma F_x = ma_x \qquad (4.3)$$
$$\Sigma F_y = ma_y$$
$$\Sigma F_z = ma_z$$

The **SI unit of force** is the newton (N), defined as the force that, when acting on a 1-kg mass, produces an acceleration of 1 m/s^2. Calculations using Equations 4.2 and 4.3 must be made using a consistent set of units for force, mass, and acceleration.

$$1\,N \equiv 1\ kg \cdot m/s^2 \qquad (4.4)$$

Weight (the gravitational force on a mass) is not an inherent property of a body, but depends on the local value of g and varies with location.

$$F_g = mg \qquad (4.5)$$

Newton's third law states that the action force, $\vec{F}_{12}$, exerted by object 1 on object 2 is equal in magnitude and opposite in direction to the reaction force, $\vec{F}_{21}$, exerted by object 2 on object 1. *Remember, the two forces in an action-reaction pair always act on two different objects—they cannot add to give a net force of zero.*

$$\vec{F}_{12} = -\vec{F}_{21} \qquad (4.6)$$

For an object in **equilibrium**, the vector sum of all the forces (the net force) acting on the object is zero. In a two-dimensional problem, the sum of all the external forces in the x and y directions must separately equal zero.

$$\sum \vec{F} = 0 \qquad (4.7)$$
$$\sum F_x = 0 \quad \sum F_y = 0 \qquad (4.8)$$

SUGGESTIONS, SKILLS, AND STRATEGIES

The following procedure is recommended when dealing with problems involving the application of Newton's second law:

- Draw a simple, neat diagram of the situation.

- Isolate the object of interest whose motion is being analyzed. Draw a free-body diagram for this object; that is, **a diagram showing all external forces acting on the object**. For systems containing more than one object, **draw separate diagrams for each object**. Do not include forces that the object exerts on its surroundings.

- Establish convenient coordinate axes for each object and find the components of the forces along these axes.

- Apply Newton's second law $\sum \vec{\mathbf{F}} = m\vec{\mathbf{a}}$, in the x and y directions for each object under consideration. For cases when the object is in equilibrium along either direction, $\sum \vec{\mathbf{F}} = 0$ along that direction.

- Solve the component equations for the unknowns. Remember that you must have as many independent equations as you have unknowns in order to obtain a complete solution.

- Often in solving such problems, one must also use the equations of kinematics (motion with constant acceleration) to find all the unknowns.

REVIEW CHECKLIST

✓ State in your own words a description of Newton's laws of motion, recall physical examples of each law, and identify the action-reaction force pairs in a multiple-body interaction problem as specified by Newton's third law.

✓ Apply Newton's laws of motion to various mechanical systems using the recommended procedure discussed in Section 4.7. Most important, you should identify all external forces acting on the system, draw the correct free-body diagrams that apply to each body of the system, and apply Newton's second law, in component form.

✓   Apply the equations of kinematics (which involve the quantities displacement, velocity, and acceleration) as described in Chapter 2 along with those methods and equations of Chapter 4 (involving mass, force, and acceleration) to the solutions of problems where both the kinematic and dynamic aspects are present.

✓   Be familiar with solving several linear equations simultaneously for the unknown quantities. *Remember, you must have as many independent equations as you have unknowns.*

ANSWERS TO SELECTED QUESTIONS

Q4.3   In the motion picture *It Happened One Night* (Columbia Pictures, 1934), Clark Gable is standing inside a stationary bus in front of Claudette Colbert, who is seated. The bus suddenly starts moving forward and Clark falls into Claudette's lap. Why did this happen?

Answer   When the bus starts moving, the mass of Claudette is accelerated by the force exerted by the back of the seat on her body. Clark is standing, however, and the only force on him is the friction between his shoes and the floor of the bus. Thus, when the bus starts moving, his feet start accelerating forward, but the rest of his body experiences almost no accelerating force (only that due to his being attached to his accelerating feet!). As a consequence, his body tends to stay almost at rest, according to Newton's first law, relative to the ground. Relative to Claudette, however, he is moving toward her and falls into her lap. (Both performers won Academy Awards.)

Q4.7   A rubber ball is dropped onto the floor. What force causes the ball to bounce?

Answer   When the ball hits the floor, it is compressed. As the ball returns to its original shape, it exerts a force on the floor, and the reaction to this thrusts it back into the air.

Q4.11   A weightlifter stands on a bathroom scale. He pumps a barbell up and down. What happens to the reading on the bathroom scale as he does so? What if he is strong enough to actually throw the barbell upward? How does the reading on the scale vary now?

continued on next page

Answer If the barbell is not moving, the reading on the bathroom scale is the combined weight of the weightlifter and the barbell. At the beginning of the lift of the barbell, the barbell accelerates upward. By Newton's third law, the barbell pushes downward on the hands of the weightlifter with more force than its weight, in order to accelerate. As a result, he is pushed with more force into the scale, increasing its reading. Near the top of the lift, the weightlifter reduces the upward force, so that the acceleration of the barbell is downward, causing it to come to rest. While the barbell is coming to rest, it pushes with less force on the weightlifter's hands, so the reading on the scale is below the combined stationary weight. If the barbell is held at rest for a moment at the top of the lift, the scale reading is simply the combined weight. As it begins to be brought down, the reading decreases, as the force of the weightlifter on the barbell is reduced. The reading increases as the barbell is slowed down at the bottom.

If we now consider the throwing of the barbell, we have the same behavior as before, except that the variations in scale reading will be larger, since more force must be applied to throw the barbell upward rather than just lift it. Once the barbell leaves the weightlifter's hands, the reading will suddenly drop to just the weight of the weightlifter, and will rise suddenly when the barbell is caught.

SOLUTIONS TO SELECTED PROBLEMS

P4.5 To model a spacecraft, a toy rocket engine is securely fastened to a large puck, which can glide with negligible friction over a horizontal surface, taken as the *xy* plane. The 4.00-kg puck has a velocity of $3.00\hat{i}$ m/s at one instant. Eight seconds later, its velocity is to be $(8.00\hat{i} + 10.0\hat{j})$ m/s. Assuming the rocket engine exerts a constant horizontal force,

(a) Find the components of the force.

(b) Find its magnitude.

Solution We use the particle under constant acceleration and particle under net force models. We first calculate the acceleration of the puck:

$$\vec{a} = \frac{\Delta\vec{v}}{\Delta t} = \frac{(8.00\hat{i} + 10.0\hat{j}) \text{ m/s} - 3.00\hat{i} \text{ m/s}}{8.00 \text{ s}} = 0.625\hat{i} \text{ m/s}^2 + 1.25\hat{j} \text{ m/s}^2$$

In $\sum\vec{F} = m\vec{a}$, the only horizontal force is the thrust $\vec{F}$ of the rocket:

(a) $\vec{F} = (4.00 \text{ kg})(0.625\hat{i} \text{ m/s}^2 + 1.25\hat{j} \text{ m/s}^2) = 2.50\hat{i} \text{ N} + 5.00\hat{j} \text{ N}$ ◊

(b) Then, $|\vec{F}| = \sqrt{(2.50 \text{ N})^2 + (5.00 \text{ N})^2} = 5.59 \text{ N}$ ◊

P4.7 Two forces, $\vec{F}_1$ and $\vec{F}_2$, act on a 5.00-kg object. If $F_1 = 20.0$ N and $F_2 = 15.0$ N, find the accelerations in (a) and (b) of Figure P4.7.

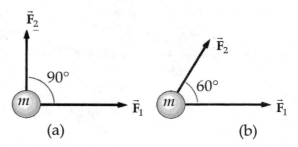

FIG. P4.7

Solution We use the particle under a net force model:

$$m = 5.00 \text{ kg}$$

(a) $$\sum \vec{F} = \vec{F}_1 + \vec{F}_2 = \left(20.0\hat{i} + 15.0\hat{j}\right) \text{ N}$$

$$\vec{a} = \frac{\sum \vec{F}}{m} = \left(4.00\hat{i} + 3.00\hat{j}\right) \text{ m/s}^2 = 5.00 \text{ m/s}^2 \text{ at } 36.9° \qquad \lozenge$$

(b) $$\sum \vec{F} = \vec{F}_1 + \vec{F}_2 = \left[20.0\hat{i} + \left(15.0\cos 60°\hat{i} + 15.0\sin 60°\hat{j}\right)\right] \text{ N} = \left(27.5\hat{i} + 13.0\hat{j}\right) \text{ N}$$

$$\vec{a} = \frac{\sum \vec{F}}{m} = \left(5.50\hat{i} + 2.60\hat{j}\right) \text{ m/s}^2 = 6.08 \text{ m/s}^2 \text{ at } 25.3° \qquad \lozenge$$

P4.13 An electron of mass 9.11×10^{-31} kg has an initial speed of 3.00×10^5 m/s. It travels in a straight line, and its speed increases to 7.00×10^5 m/s in a distance of 5.00 cm. Assuming its acceleration is constant,

(a) Determine the total force on the electron.

(b) Compare this force with the weight of the electron.

Solution **Conceptualize:** We should expect that only a very small force is required to accelerate an electron because of its small mass, but this force is probably much greater than the weight of the electron if the gravitational force can be neglected.

Categorize: Since this is simply a linear acceleration problem, we can use Newton's second law to find the force as long as the electron does not approach relativistic speeds (much less than 3×10^8 m/s), which is certainly the case for this problem. We know the initial and final velocities, and the distance involved, so from these we can find the acceleration needed to determine the force.

continued on next page

Analyze: From $v_f^2 = v_i^2 + 2ax$ and $\sum F = ma$ we can solve for the acceleration and the force $a = \dfrac{v_f^2 - v_i^2}{2x}$ and so $\sum F = \dfrac{m\left(v_f^2 - v_i^2\right)}{2x}$.

(a) $\sum F = \dfrac{\left(9.11 \times 10^{-31} \text{ kg}\right)\left(\left(7.00 \times 10^5 \text{ m/s}\right)^2 - \left(3.00 \times 10^5 \text{ m/s}\right)^2\right)}{(2)(0.0500 \text{ m})}$

$\sum F = 3.64 \times 10^{-18}$ N　　　　　　　　　　　　　　　　◊

(b) The weight of the electron is

$$F_g = mg = \left(9.11 \times 10^{-31} \text{ kg}\right)\left(9.80 \text{ m/s}^2\right) = 8.93 \times 10^{-30} \text{ N} \qquad ◊$$

The ratio of the accelerating force to the weight is

$$\frac{F}{F_g} = 4.08 \times 10^{11}. \qquad ◊$$

Finalize: The force that cause the electron to accelerate is indeed a small fraction of a newton, but it is much greater than the gravitational force. For this reason, it is quite reasonable to ignore the weight of the electron in problems about electric forces.

P4.19 A bag of cement of weight F_g hangs from three wires as shown in Figure P4.19. Two of the wires make angles θ_1 and θ_2 with the horizontal. If the system is in equilibrium, show that the tension in the left-hand wire is

$$T_1 = \frac{F_g \cos\theta_2}{\sin(\theta_1 + \theta_2)}$$

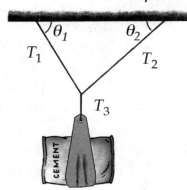

Solution We use the particle in equilibrium model. Draw a free-body diagram for the knot where the three ropes are joined. Choose the x axis to be horizontal and apply Newton's second law in component form.

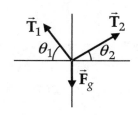

FIG. P4.19

$$\Sigma F_x = 0; \quad T_2 \cos\theta_2 - T_1 \cos\theta_1 = 0 \quad\quad (1)$$

$$\Sigma F_y = 0; \quad T_2 \sin\theta_2 + T_1 \sin\theta_1 - F_g = 0 \quad\quad (2)$$

Solve equation (1) for $\quad T_2 = \dfrac{T_1 \cos\theta_1}{\cos\theta_2}.$

Substitute this expression for T_2 into Equation (2):

$$\left(\frac{T_1 \cos\theta_1}{\cos\theta_2}\right)\sin\theta_2 + T_1\sin\theta_1 = F_g.$$

Solve for $T_1 = \dfrac{F_g \cos\theta_2}{\cos\theta_1 \sin\theta_2 + \sin\theta_1 \cos\theta_2}$. Use trigonometric identity,

$$\sin(\theta_1 + \theta_2) = \cos\theta_1 \sin\theta_2 + \sin\theta_1 \cos\theta_2$$

to find $T_1 = \dfrac{F_g \cos\theta_2}{\sin(\theta_1 + \theta_2)}.$ ◊

The equation indicates that the tension is directly proportional to the weight of the bag. As $\theta_1 + \theta_2$ approaches zero (as the angle between the two upper ropes approaches 180°) the tension goes to infinity. Making the right-hand rope horizontal maximizes the tension in the left-hand rope, according to the proportionality of T_1 to $\cos\theta_2$. If the right-hand rope is vertical, the tension in the left-hand rope is zero.

P4.27   A 1.00-kg object is observed to accelerate at 10.0 m/s^2 in a direction 30.0° north of east (Fig. P4.27). The force $\vec{F}_2$ acting on the mass has magnitude 5.00 N and is directed north. Determine the magnitude and direction of the force $\vec{F}_1$ acting on the mass.

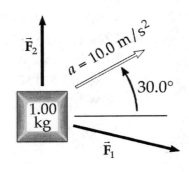

FIG. P4.27

Solution   Conceptualize: The net force acting on the mass is $\sum F = ma = (1 \text{ kg})(10 \text{ m/s}^2) = 10$ N. If we sketch a vector diagram of the forces drawn to scale $(\vec{F}_2 = \sum\vec{F} - \vec{F}_1)$, we see that $F_1 \approx 9$ N, to the east.

Categorize: We can find a more precise result by examining the forces in terms of vector components. For convenience, we choose directions east and north along $\hat{i}$ and $\hat{j}$, respectively.

Analyze: $\vec{a} = [(10.0\cos30.0°)\hat{i} + (10.0\sin30.0°)\hat{j}] \text{ m/s}^2 = (8.66\hat{i} + 5.00\hat{j}) \text{ m/s}^2$
From Newton's second law,

$$\sum\vec{F} = m\vec{a} = (1.00 \text{ kg})(8.66\hat{i} \text{ m/s}^2 + 5.00\hat{j} \text{ m/s}^2) = (8.66\hat{i} + 5.00\hat{j}) \text{ N}$$

and $\sum\vec{F} = \vec{F}_1 + \vec{F}_2$

so $\vec{F}_1 = \sum\vec{F} - \vec{F}_2 = (8.66\hat{i} + 5.00\hat{j} - 5.00\hat{j}) \text{ N} = 8.66\hat{i} \text{ N} = 8.66 \text{ N east}.$ ◊

Finalize: Our calculated answer agrees with the prediction from the force diagram.

P4.29   A block is given an initial velocity of 5.00 m/s up a frictionless 20.0° incline (see Figure P4.29). How far up the incline does the block slide before coming to rest?

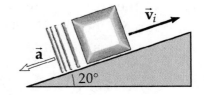

Solution   Every successful physics student (this means you) learns to solve inclined-plane problems.

Hint one: Try to set the axes in the direction of motion. In this case, take the x-axis along the incline, so that $a_y = 0$.

Hint two: Recognize that the 20.0° angle between the x-axis and horizontal implies a 20.0° angle between the weight vector and the y-axis. Why? Because "angles are equal if their sides are perpendicular, right side to right side and left side to left side." Either you learned this theorem in geometry class, or you learn it now, since it is a theorem used often in physics.

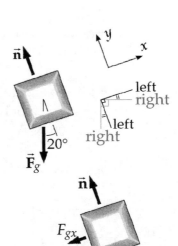

Hint three: The 20.0° angle lies between $m\vec{g}$ and the y-axis, so split the weight vector into its x and y components:

FIG. P4.29

$$mg_x = -mg\sin 20.0° \qquad mg_y = -mg\cos 20.0°$$

Now, Newton's law applies for each axis. Applying it to the x axis, $\Sigma F_x = ma_x$:
$-mg\sin 20° = ma_x$

$$a_x = -g\sin 20° = -(9.80 \text{ m/s}^2)\sin 20° = -3.35 \text{ m/s}^2 .$$

From Eq. 2.13, $v_{xf}^{\ 2} = v_{xi}^{\ 2} + 2a_x(x_f - x_i)$

$$0 = (5.00 \text{ m/s})^2 + 2(-3.35 \text{ m/s}^2)(x_f - x_i).$$

Solving, $(x_f - x_i) = 3.73 \text{ m}.$ ◊

P4.35   In the system shown in Figure P4.35, a horizontal
force $\vec{F}_x$ acts on the 8.00-kg mass. The horizontal
surface is frictionless.

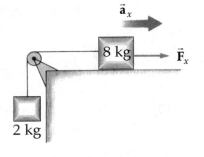

(a)   For what values of F_x does the 2.00-kg mass
accelerate upward?

(b)   For what values of F_x is the tension in the
cord zero?

(c)   Plot the acceleration of the 8.00-kg object
versus F_x. Include values of F_x from –100 N
to +100 N.

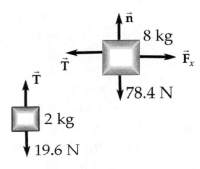

Solution The blocks' weights are:

FIG. P4.35

$$F_{g1} = m_1 g = (8.00 \text{ kg})(9.80 \text{ m/s}^2) = 78.4 \text{ N}$$

$$F_{g2} = m_2 g = (2.00 \text{ kg})(9.80 \text{ m/s}^2) = 19.6 \text{ N}$$

Let T be the tension in the connecting cord and draw a free-body diagram for
each block.

(a)   For the 2-kg mass, with the y-axis directed upwards,

$$\Sigma F_y = m a_y \qquad \text{yields} \qquad T - 19.6 \text{ N} = (2.00 \text{ kg})a_y \qquad (1)$$

Thus, we find that $a_y > 0$ when $T > 19.6$ N. For acceleration by the system
of two blocks, $F_x \geq T$, so $F_x > 19.6$ N whenever the 2-kg mass accelerates
upward.   ◊

(b)   Looking at the free-body diagram for the 8.00-kg mass, and taking the $+x$
direction to be directed to the right, we can apply Newton's law in the
horizontal direction:

From $\Sigma F_x = m a_x$ $-T + F_x = (8.00 \text{ kg})a_x$ $\qquad (2)$

If $T = 0$, the cord goes slack; the 2-kg object is in free fall. The 8-kg object
can have an acceleration to the left of larger magnitude:

$$a_x \leq -9.80 \text{ m/s}^2 \text{ with } F_x \leq -78.4 \text{ N}.$$ ◊

continued on next page

(c)   If $F_x \geq -78.4$ N, then both equations (1) and (2) apply. Substituting the value for T from the (1) into (2),

$$-(2.00 \text{ kg})a_y - 19.6 \text{ N} + F_x = (8.00 \text{ kg})a_x$$

In this case $a_x = a_y$: $F_x = (8.00 \text{ kg} + 2.00 \text{ kg})a_x + 19.6$ N

$$a_x = \frac{F_x}{10.0 \text{ kg}} - 1.96 \text{ m/s}^2 \qquad\qquad (F_x \geq -78.4 \text{ N}) \qquad\qquad (3)$$

From part (b), we find that if $F_x \leq -78.4$ N, then $T = 0$ and equation (2) becomes $F_x = (8.00 \text{ kg})a_x$:

$$a_x = \frac{F_x}{8.00 \text{ kg}} \qquad\qquad (F_x \leq -78.4 \text{ N}) \qquad\qquad (4)$$

Observe that we have translated the pictorial representation into a simplified pictorial representation and then into a mathematical representation. We proceed to a tabular representation and a graphical representation of equations (3) and (4):

F_x	a_x
−100 N	−12.5 m/s²
−50.0 N	−6.96 m/s²
0	−1.96 m/s²
50.0 N	3.04 m/s²
100 N	8.04 m/s²
150 N	13.04 m/s²

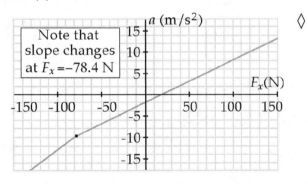

FIG. P4.35(c)

P4.37   A 72.0-kg man stands on a spring scale in an elevator. Starting from rest, the elevator ascends, attaining its maximum speed of 1.20 m/s in 0.800 s. It travels with this constant speed for the next 5.00 s. The elevator then undergoes a uniform acceleration in the negative y direction for 1.50 s and comes to rest. What does the spring scale register

(a)   before the elevator starts to move?

(b)   during the first 0.800 s?

continued on next page

(c)   while the elevator is traveling at constant speed?

(d)   during the time it is slowing down?

Solution Conceptualize: Based on sensations experienced riding in an elevator, we expect that the man should feel slightly heavier when the elevator first starts to ascend, lighter when it comes to a stop, and his normal weight when the elevator is not accelerating. His apparent weight is registered by the spring scale beneath his feet, so the scale force should correspond to the force he feels through his legs (Newton's third law).

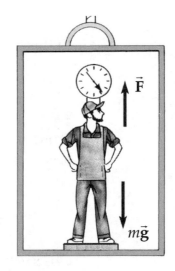

Categorize: We should draw free body diagrams for each part of the elevator trip and apply Newton's second law to find the scale force. The acceleration can be found from the change in speed divided by the elapsed time.

Analyze: Consider the free-body diagram of the man shown to the right. The force F is the upward force exerted on the man by the scale, and his weight is

FIG. P4.37

$$F_g = mg = (72.0 \text{ kg})(9.80 \text{ m/s}^2) = 706 \text{ N}.$$

With $+y$ defined to be upwards, Newton's 2nd law gives $\sum F_y = +F_s - F_g = ma$. Thus, we calculate the upward scale force to be

$$F_s = 706 \text{ N} + (72.0 \text{ kg})a \qquad\qquad (1)$$

where a is the acceleration the man experiences as the elevator changes speed.

(a)   Before the elevator starts moving, the elevator's acceleration is zero ($a = 0$). Therefore, Equation (1) gives the force exerted by the scale on the man as 706 N (upward), and the man exerts a downward force of 706 N on the scale.   ◊

continued on next page

(b)   During the first 0.800 s of motion, the man accelerates at a rate of

$$a = \frac{\Delta v}{\Delta t} = \frac{1.20 \text{ m/s} - 0}{0.800 \text{ s}} = 1.50 \text{ m/s}^2$$

Substituting a into Eq. (1) then gives:

$$F = 706 \text{ N} + (72.0 \text{ kg})(1.50 \text{ m/s}^2) = 814 \text{ N} \qquad \lozenge$$

(c)   While the elevator is traveling upward at a constant speed, the acceleration is zero and Equation (1) again gives a scale force $F = 706 \text{ N}$. $\lozenge$

(d)   During the last 1.50 s, the elevator starts with an upward velocity of 1.20 m/s, and comes to rest with an acceleration of

$$a = \frac{\Delta v}{\Delta t} = \frac{0 - 1.20 \text{ m/s}}{1.50 \text{ s}} = -0.800 \text{ m/s}^2$$

Thus, the force of the man on the scale is:

$$F = 706 \text{ N} + (72.0 \text{ kg})(-0.800 \text{ m/s}^2) = 648 \text{ N} \qquad \lozenge$$

Finalize: The calculated scale forces are consistent with our predictions. This problem could be extended to a couple of extreme cases. If the acceleration of the elevator were $+9.80 \text{ m/s}^2$, then the man would feel twice as heavy, and if $a = -9.80 \text{ m/s}^2$ (free fall), then he would feel "weightless", even though his true weight $\left(F_g = mg\right)$ would remain the same.

P4.41   An inventive child named Pat wants to reach an apple in a tree without climbing the tree. Sitting in a chair connected to a rope that passes over a frictionless pulley (Fig. P4.41), Pat pulls on the loose end of the rope with such a force that the spring scale reads 250 N. Pat's true weight is 320 N, and the chair weighs 160 N.

   (a)   Draw free-body diagrams for Pat and the chair considered as separate systems and another diagram for Pat and the chair considered as one system.

   (b)   Show that the acceleration of the system is upward and find its magnitude.

   (c)   Find the force Pat exerts on the chair.

FIG. P4.41

Solution   (a)

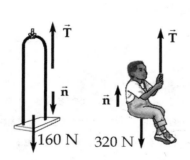

   (b)   First consider Pat and the chair as the system. Note that two ropes support the system, and $T = 250$ N in each rope.

   Applying $\Sigma F = ma$, $2T - (160 \text{ N} + 320 \text{ N}) = ma$

   where $m = \dfrac{480 \text{ N}}{9.80 \text{ m/s}^2} = 49.0$ kg.

   Solving for a gives $a = \dfrac{(500 - 480) \text{ N}}{49.0 \text{ kg}} = 0.408 \text{ m/s}^2.$ ◊

continued on next page

(c)   On Pat, we apply $\Sigma F = ma$: $n + T - 320\text{ N} = ma$

where

$$m = \frac{320\text{ N}}{9.80\text{ m/s}^2} = 32.7\text{ kg}$$

$$n = ma + 320\text{ N} - T$$

$$n = (32.7\text{ kg})(0.408\text{ m/s}^2) + 320\text{ N} - 250\text{ N}$$

Therefore, $n = 83.3\text{ N}$.   ◊

P4.45   An object of mass M is held in place by an applied force $\vec{F}$ and a pulley system as shown in Figure P4.45. The pulleys are massless and frictionless.

(a)   Find the tension in each section of rope, T_1, T_2, T_3, T_4, and T_5.

(b)   Find the magnitude of $\vec{F}$. *Suggestion*: draw a free-body diagram for each pulley.

Solution (a)   Draw free-body diagrams and apply Newton's 2nd law. (All forces are along the y axis.)

For M,

$$T_5 = Mg$$

Assume frictionless pulleys. The tension is constant throughout a light, continuous rope.

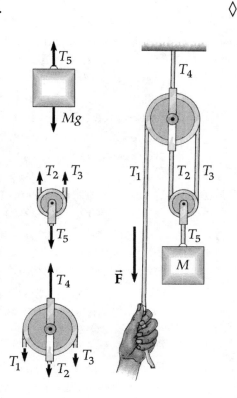

FIG. P4.45

Therefore, $T_1 = T_2 = T_3$.

For the bottom pulley, $\Sigma F = 0 = T_2 + T_3 - T_5$.

So $2T_2 = T_5$

and $T_1 = T_2 = T_3 = \dfrac{1}{2}Mg$.   ◊

continued on next page

(b)   The applied force is $F = T_1 = \dfrac{1}{2} Mg.$ ◊

For the top pulley, $\sum F = 0 = T_4 - T_1 - T_2 - T_3.$

Solving, $T_4 = T_1 + T_2 + T_3 = \dfrac{3}{2} Mg.$ ◊

P4.47   What horizontal force must be applied to the cart shown in Figure P4.47 in order that the blocks remain stationary relative to the cart? Assume all surfaces, wheels, and pulley are frictionless. *Suggestion:* Note that the force exerted by the string accelerates m_1.

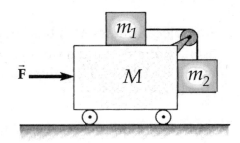

Solution Draw separate free-body diagrams for blocks m_1 and m_2. Remembering that normal forces are always perpendicular to the contacting surface, and always push on a body, draw $\vec{n}_1$ and $\vec{n}_2$ as shown. Note that m_2 should be in contact with the cart, and therefore does have a normal force from the cart. Remembering that ropes always pull on bodies in the direction of the rope, draw the tension force $\vec{T}$. Finally, draw the gravitational force on each block, which always points downwards.

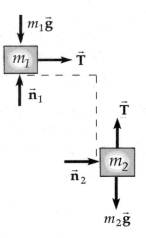

FIG. P4.47

Conceptualize: What can keep m_2 from falling? Only tension in the cord connecting it with m_1. This tension pulls forward on m_1 to accelerate that mass. We might guess that the acceleration is proportional to both m_2 and g and inversely proportional to m_1, so perhaps $a = \dfrac{m_2 g}{m_1}$. If the entire system accelerates at this rate, then m_1 need not slide on M to achieve this acceleration. We should also expect the applied force to be proportional to the total mass of the system.

continued on next page

Catgeorize: Use $\sum F = ma$ and the free-body diagrams above.

Analyze: For m_2, $\qquad\qquad T - m_2 g = 0 \quad$ or $\quad T = m_2 g$

For m_1, $\qquad\qquad T = m_1 a \quad$ or $\quad a = \dfrac{T}{m_1}$

Substituting for T, we have $\qquad a = \dfrac{m_2 g}{m_1}$

For all 3 blocks, $\qquad F = (M + m_1 + m_2)a$

Therefore, $\qquad F = (M + m_1 + m_2)\left(\dfrac{m_2 g}{m_1}\right)$ $\qquad\qquad\qquad\Diamond$

Finalize: Even though this problem did not have a numerical solution, we were still able to rationalize the algebraic form of the solution. This technique does not always work, especially for complex situations, but often we can think through a problem to see if an equation for the solution makes sense based on the physical principles we know.

P4.53 A van accelerates down a hill (Fig. P4.53), going from rest to 30.0 m/s in 6.00 s. During the acceleration, a toy ($m = 0.100$ kg) hangs by a string from the van's ceiling. The acceleration is such that the string remains perpendicular to the ceiling.

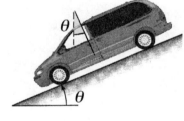

(a) Determine the angle θ.

(b) Determine the tension in the string.

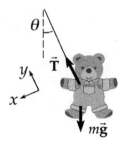

Solution The acceleration is obtained from $v_f = v_i + at$:

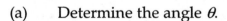

$$30.0 \text{ m/s} = 0 + a(6.00 \text{ s})$$
$$a = 5.00 \text{ m/s}^2$$

FIG. P4.53

continued on next page

The toy moves with the same acceleration as the van, 5.00 m/s^2 parallel to the hill. We take the x axis in this direction, so $a_x = 5.00$ m/s^2 and $a_y = 0$. The only forces on the toy are the string tension in the y direction and its weight, as shown in the free-body diagram.

$$mg = (0.100 \text{ kg})(9.80 \text{ } m/s^2) = 0.980 \text{ N}.$$

(a)   Using $\sum F_x = ma_x$: $(0.980 \text{ N})\sin\theta = (0.100 \text{ kg})(5.00 \text{ } m/s^2)$

$$\sin\theta = \frac{0.500}{0.980} \text{ and } \theta = 30.7° \qquad\qquad ◊$$

(b)   Using $\sum F_y = ma_y$: $+T - (0.980 \text{ N})\cos\theta = 0$

$$T = (0.980 \text{ N})\cos 30.7° = 0.843 \text{ N} \qquad\qquad ◊$$

Chapter 5

More Applications of Newton's Laws

Section 5.1   Forces of Friction

When an object is in motion either on a surface or through a viscous medium such as air or water, the object reacts with the surface or the medium through which it is moving. *The resulting resistance is called a force of friction.*

In addition, if an external force is applied to an object at rest on a rough surface such that the force has a component directed parallel to the surface, there will be an opposing force of friction characteristic of the pair of surfaces. When there is no relative motion between the object and the surface on which it rests, the force is called **static friction**. If relative motion occurs, the force is then one of **kinetic friction**.

Values of the maximum value of the force of static friction $\left(f_{s,\,max}\right)$ and the force of kinetic friction $\left(f_k\right)$ can be determined by experiment. *Both forms of frictional force are proportional to the normal force between the two surfaces.* That is,

$$f_{s,\,max} = \mu_s n \text{ and } f_k = \mu_k n.$$

It should be noted that in general, for any pair of surfaces, μ_k is generally less than μ_s. *The coefficients of friction are nearly independent of the area of contact between the surfaces.*

Section 5.2   Newton's Second Law Applied to a Particle in Uniform Circular Motion

If a particle moves in a circle of radius r with constant speed, it undergoes a centripetal acceleration $\dfrac{v^2}{r}$ directed toward the center of rotation. (Recall that in this case, the centripetal acceleration arises from the change in direction of the velocity vector.) Newton's second law applied to the motion says that the centripetal acceleration arises from some external, centripetal force acting toward the center of rotation.

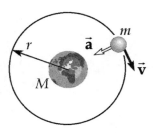

FIG. 5.1

The **universal gravitational constant**, G, is not to be confused with the acceleration due to gravity. The gravitational force is always a force of attraction and, as shown in Figure 5.2, the force on m_1 due to m_2 is equal and opposite the force on m_2 due to m_1 (Newton's third law).

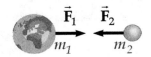

FIG. 5.2

The **gravitational force of attraction on a mass** m near the surface of the Earth (Figure 5.3) is inversely proportional to the square of the distance between the object and the center of the Earth. *The gravitational force exerted by a spherically symmetric mass distribution on a particle outside the sphere is the same as if the entire mass of the sphere were concentrated at its center.*

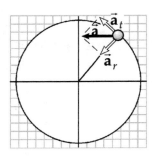

FIG. 5.3

Section 5.3   Nonuniform Circular Motion

The total acceleration of a particle in non-uniform circular motion is the vector sum of the centripetal acceleration $\vec{a}_r$ (directed toward the center of the path and produces a change in the direction of the velocity vector) and the tangential acceleration $\vec{a}_t$ (directed along the tangent to the path and produces a change in the magnitude of the velocity vector). As illustrated in Figure 5.4, $\vec{a} = \vec{a}_r + \vec{a}_t$.

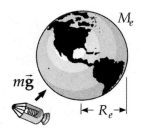

FIG. 5.4

Section 5.4   Motion in the Presence of Velocity-Dependent Resistive Forces

A body moving through a gas or liquid experiences a resistive force (opposing its motion), which can have a complicated velocity dependence. A falling body reaches a terminal velocity (maximum velocity) when the downward force of gravity is balanced by the upward resistive force. That is, when $\sum \vec{F} = 0$, $\vec{a} = 0$, and $\vec{v} = $ constant.

Section 5.5   The Fundamental Forces of Nature

The **gravitational force** is the mutual force of attraction between any two objects in the universe; it is the weakest of the fundamental forces.

The **electromagnetic force** involves two types of particles: those with positive charge, and those with negative charge. It is this force that binds atoms and molecules to form matter. The electromagnetic force is stronger than the gravitational force.

The **strong force** is a very short range force that binds the nucleons to form a nucleus.

The **weak force** plays a role in radioactive decay and is much stronger than the gravitational force but much weaker than the electromagnetic force.

It is now known that the electromagnetic force and the weak force are both manifestations of a single force called the **electro-weak force**.

EQUATIONS AND CONCEPTS

The **force of static friction** between two surfaces in contact but not in motion, relative to each other, cannot be greater than $\mu_s n$, where n is the normal (perpendicular) force between the two surfaces and μ_s (coefficient of static friction) is a dimensionless constant that depends on the nature of the pair of surfaces.

$$f_s \leq \mu_s n \qquad (5.1)$$

The **force of kinetic friction** applies when two surfaces are in relative motion. *The friction force is parallel to the surface on which an object is in contact and is directed opposite the direction of actual or impending motion.*

$$f_k = \mu_k n \qquad (5.2)$$

When an object is in **uniform circular motion**, the net force acting on the object is a centripetal force (directed toward the center of the circular path.)

$$\Sigma F_r = m\,a_c = m\frac{v^2}{r} \qquad (5.3)$$

A **resistive force** $\vec{R}$ will be exerted on an object moving through a medium (gas or liquid). The form of Equation 5.4 assumes that the resistive force is proportional to the speed of the object. *The constant b has a value that depends on the properties of the medium and the dimensions and shape of the object.*

$$\vec{R} = -b\vec{v} \qquad (5.4)$$

A **differential equation** is used to describe the motion of an object falling vertically in a viscous medium. If we consider a special case in which the resistive force is proportional to the velocity, $R = bv$, then the differential equation has the form of Equation 5.5.

$$\frac{dv}{dt} = g - \frac{b}{m}v \qquad (5.5)$$

or

$$mg - bv = m\frac{dv}{dt}$$

The **speed as a function of time** is found using Equation 5.6, when the object is released from rest at $t = 0$.

$$v = v_T\left(1 - e^{-t/\tau}\right) \qquad (5.6)$$

An object reaches **terminal speed** as the magnitude of the resistive force approaches the weight of the object.

$$v_T = \frac{mg}{b}$$

The **time constant** τ is the time at which an object, released from rest, will achieve a speed equal to 63.2% of the terminal speed.

$$\tau = \frac{m}{b}$$

For large **objects moving at high speed**, the resistive force is modeled as being proportional to v^2.

$$R = \frac{1}{2}D\rho A v^2 \qquad (5.7)$$

ρ = density of the air
A = cross-sectional area of the object
D = drag coefficient (dimensionless, empirical quantity)

Terminal speed v_T is achieved when the gravitational force is balanced by the resistive force; the net force is zero and therefore the acceleration is zero.

$$v_T = \sqrt{\frac{2mg}{D\rho A}} \qquad (5.10)$$

Newton's law of gravitation states that every particle in the universe attracts every other particle with a force directly proportional to the product of the two masses and inversely proportional to the square of the distance between them.

$$F_g = G\frac{m_1 m_2}{r^2} \qquad (5.11)$$

Coulomb's law expresses the magnitude of the electrostatic force between two charged particles separated by a distance r. Opposite sign charges attract each other, and like sign charges repel.

$$F_e = k_e\frac{q_1 q_2}{r^2} \qquad (5.12)$$

SUGGESTIONS, SKILLS, AND STRATEGIES

Section 5.4 deals with the motion of a body through a gas or liquid. If you covered this section in class, the following solution to Equation 5.5 (when the resistive force $\vec{\mathbf{R}} = -b\vec{\mathbf{v}}$) may be useful to know:

$$\frac{dv}{dt} = g - \frac{b}{m}v \qquad (5.5)$$

In order to solve this equation, it helps to change variables. If we let $y = g - \left(\dfrac{b}{m}\right)v$, it follows that $dy = -\left(\dfrac{b}{m}\right)dv$. With these substitutions, Equation 5.5 becomes

$$-\left(\frac{m}{b}\right)\frac{dy}{dt} = y \ \text{ or } \ \frac{dy}{y} = -\frac{b}{m}dt \ .$$

Integrating this expression (now that the variables are separated) gives

$$\int\frac{dy}{y} = -\frac{b}{m}\int dt \ \text{ or } \ \ln y = -\frac{b}{m}t + \text{const.}$$

This is equivalent to $y = (\text{const})\, e^{-bt/m} = g - \dfrac{b}{m}v$. Taking $v = 0$ at $t = 0$, we see $\text{const} = g$.

so $v = \dfrac{mg}{b}(1 - e^{-bt/m}) = v_t(1 - e^{-t/\tau})$ (5.6)

where $\tau = \dfrac{m}{b}.$

REVIEW CHECKLIST

✓   Discuss Newton's universal law of gravity (the inverse-square law), and understand that it is an attractive force between two particles separated by a distance r.

✓   Discuss the nature of the fundamental forces in nature (gravitational, electromagnetic, weak, and strong) and characterize the properties and relative strengths of these forces.

✓   Apply Newton's second law to uniform and nonuniform circular motion.

✓   Recognize that motion of an object through a liquid or gas can involve resistive forces that have a complicated velocity dependence.

ANSWERS TO SELECTED QUESTIONS

Q5.3   Identify the action-reaction pairs in the following situations: a man takes a step; a snowball hits a girl in the back; a baseball player catches a ball; a gust of wind strikes a window.

Answer   As a man takes a step, the action is the force his foot exerts on the Earth; the reaction is the force of the Earth on his foot. In the second case, the action is the force exerted on the girl's back by the snowball; the reaction is the force exerted on the snowball by the girl's back. The third action is the force of the glove on the ball; the reaction is the force of the ball on the glove. The fourth action is the force exerted on the window by the air molecules; the reaction is the force on the air molecules exerted by the window. We could equally well interchange the terms 'action' and 'reaction' in each case.

Q5.5   Suppose you are driving a classic car. Why should you avoid slamming on your brakes if you want to stop in the shortest possible distance? (Many cars have antilock brakes that avoid this problem.)

Answer   The brakes may lock and the car will slide farther since the coefficient of sliding friction is less than the coefficient of static friction. If the wheels continue to roll, the force of static friction will make the car slow down.

Q5.10   It has been suggested that rotating cylinders about 10 mi in length and 5 mi in diameter be placed in space and used as colonies. The purpose of the rotation is to simulate gravity for the inhabitants. Explain this concept for producing an effective imitation of gravity.

Answer   The centripetal force on the inhabitants is provided by the normal force exerted on them by the cylinder wall. If the rotation rate is adjusted to such a speed that this normal force is equal to their weight on Earth, the inhabitants would not be able to distinguish between this artificial gravity and normal gravity.

Q5.12   Why does a pilot tend to black out when pulling out of a steep dive?

Answer   When pulling out of a dive, blood leaves the pilot's head because the pilot's blood pressure is not great enough to compensate for both the gravitational force and the centripetal acceleration of the airplane. This loss of blood from the brain can cause the pilot to lose consciousness.

Q5.14   A falling sky diver reaches terminal speed with her parachute closed. After the parachute is opened, what parameters change to decrease this terminal speed?

Answer   From the expression for the force of air resistance and Newton's law, we derive the equation that governs the motion of the skydiver:

$$m\frac{dv_y}{dt} = mg - \frac{D\rho A}{2}v_y{}^2$$

where D is the coefficient of drag of the parachutist, and A is the area of the parachutist's body. At terminal speed,

$$a_y = \frac{dv_y}{dt} = 0 \qquad \text{and} \qquad v_t = \sqrt{\frac{2mg}{D\rho A}}$$

When the parachute opens, the coefficient of drag D and the effective area A both increase, thus reducing the velocity of the skydiver.

Modern parachutes also add a third term, lift, to change the equation to

$$m\frac{dv_y}{dt} = mg - \frac{D\rho A}{2}v_y{}^2 - \frac{L\rho A}{2}v_x{}^2$$

where v_y is the vertical velocity, and v_x is the horizontal velocity. This lift is best seen in the "Paraplane," an ultralight airplane made from a fan, a chair, and a parachute.

SOLUTIONS TO SELECTED PROBLEMS

P5.9   A 3.00-kg block starts from rest at the top of a 30.0° incline and slides a distance of 2.00 m down the incline in 1.50 s.

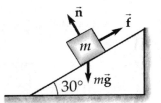

(a)   Find the magnitude of the acceleration of the block.

(b)   Find the coefficient of kinetic friction between block and plane.

FIG. P5.9

(c)   Find the friction force acting on the block.

(d)   Find the speed of the block after it has slid 2.00 m.

Solution   We use the particle under constant acceleration and particle under net force models. We choose $x_i = 0$ at the starting point of the motion.

(a)   At constant acceleration, $x_f = v_i t + \frac{1}{2}at^2$.

So, $a = \dfrac{2(x_f - v_i t)}{t^2} = \dfrac{2(2.00 \text{ m} - 0)}{(1.50 \text{ s})^2} = 1.78 \text{ m/s}^2$.   ◊

From the acceleration, we can calculate the friction force, answer (c), next.

(c)   Choose the *x* axis parallel to the incline, take the positive direction down the incline (in the direction of the acceleration) and apply the second law.

$\Sigma F_x = mg \sin \theta - f = ma$: $f = m(g \sin \theta - a)$

$f = (3.00 \text{ kg})\left[9.80 \text{ m/s}^2 (\sin 30.0°) - 1.78 \text{ m/s}^2\right] = 9.37 \text{ N}$ ◊

(b)   Applying Newton's law in the *y* direction (perpendicular to the incline),

$\sum F_y = n - mg \cos \theta = 0 : n = mg \cos \theta$.

Because $f = \mu n$, $\mu = \dfrac{f}{mg \cos \theta} = \dfrac{9.37 \text{ N}}{(3.00 \text{ kg})(9.80 \text{ m/s}^2) \cos 30.0°} = 0.368$ ◊

(d) $v_f = v_i + at$, so $v = 0 + (1.78 \text{ m/s}^2)(1.50 \text{ s}) = 2.67 \text{ m/s}$ ◊

P5.11 Two blocks connected by a rope of negligible mass are being dragged by a horizontal force $\vec{F}$ (Fig. P5.11). Suppose that $F = 68.0$ N, $m_1 = 12.0$ kg, $m_2 = 18.0$ kg, and the coefficient of kinetic friction between each block and the surface is 0.100.

FIG. P5.11

(a) Draw a free-body diagram for each block.

(b) Determine the tension, T, and the magnitude of the acceleration of the system.

Solution (a) The free-body diagrams for m_1 and m_2 are:

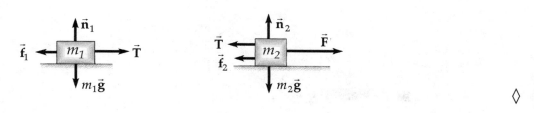

◊

(b) Use the free-body diagrams to apply Newton's second law. For m_1:
$\Sigma F_x = T - f_1 = m_1 a$ or

$$T = m_1 a + f_1 \tag{1}$$

$\Sigma F_y = n_1 - m_1 g = 0$ or $n_1 = m_1 g$.

Also, $f_1 = \mu n_1 = (0.100)(12.0 \text{ kg})(9.80 \text{ m/s}^2) = 11.8$ N.

For m_2: $\Sigma F_x = F - T - f_2 = m_2 a$ or

$$T = F - m_2 a - f_2 \tag{2}$$

$\Sigma F_y = n_2 - m_2 g = 0$ or $n_2 = m_2 g$.

Also, $f_2 = \mu n_2 = (0.100)(18.0 \text{ kg})(9.80 \text{ m/s}^2) = 17.6$ N. Substituting T from equation (1) into (2), we get $m_1 a + f_1 = F - m_2 a - f_2$.

Solving for a, $a = \dfrac{F - f_1 - f_2}{m_1 + m_2} = \dfrac{(68.0 - 11.8 - 17.6) \text{ N}}{(12.0 + 18.0) \text{ kg}} = 1.29 \text{ m/s}^2$. ◊

From Eq. (1), $T = m_1 a + f_1 = (12.0 \text{ kg})(1.29 \text{ m/s}^2) + 11.8 \text{ N} = 27.2 \text{ N}$ ◊

P5.15   A light string can support a stationary hanging load of 25.0 kg before breaking. A 3.00-kg mass attached to the light string rotates on a horizontal, frictionless table in a circle of radius 0.800 m and the other end of the string is held fixed. What range of speeds can the object have before the string breaks?

Solution   We use the particle under net force and particle in uniform circular motion models. The string will break if the tension T exceeds

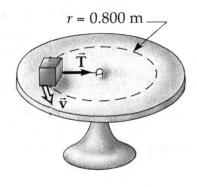

$r = 0.800$ m

$$T_{max} = mg = (25.0 \text{ kg})(9.80 \text{ m/s}^2) = 245 \text{ N}$$

As the 3.00-kg mass rotates in a horizontal circle, the tension provides the centripetal acceleration:

$$a = \frac{v^2}{r}.$$

FIG. P5.15

From $\sum F = ma$, $T = \dfrac{mv^2}{r}$.

Therefore, $v^2 = \dfrac{rT}{m} = \dfrac{(0.800 \text{ m})T}{(3.00 \text{ kg})} \le \dfrac{(0.800 \text{ m})}{(3.00 \text{ kg})} T_{max}.$

Substituting $T_{max} = 245$ N, we find $v^2 \le 65.3 \text{ m}^2/\text{s}^2$

and $0 < v < 8.08$ m/s.   ◊

P5.17   A crate of eggs is located in the middle of the flatbed of a pickup truck as the truck negotiates an unbanked curve in the road. The curve may be regarded as an arc of a circle of radius 35.0 m. If the coefficient of static friction between crate and truck is 0.600, how fast can the truck be moving without the crate sliding?

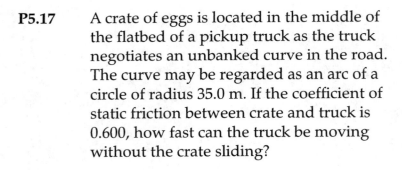
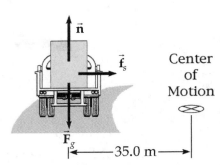

FIG. P5.17

Solution   Call the mass of the egg crate m. The forces on it are its weight $\vec{F} = m\vec{g}$ vertically down, the normal force $\vec{n}$ of the truck bed vertically up, and static friction $\vec{f}_s$ directed to oppose relative sliding motion of the crate over the truck bed. The friction force is directed radially inward. It is the only horizontal force on the crate, so it must provide the centripetal acceleration. When the truck has maximum speed, friction f_s will have its maximum value with $f_s = \mu_s n$.

continued on next page

$\Sigma F_y = ma_y$ gives $n - mg = 0$ or $n = mg$

$\Sigma F_x = ma_x$ gives $f_s = ma_r$

From these two equations, $\mu_s n = \dfrac{mv^2}{r}$ and $\mu_s mg = \dfrac{mv^2}{r}$

The mass divides out, leaving

$$v = \sqrt{\mu_s gr} = \sqrt{(0.600)(9.80 \ \text{m/s}^2)(35.0 \ \text{m})} = 14.3 \ \text{m/s} \qquad \Diamond$$

P5.21   Tarzan ($m = 85.0$ kg) tries to cross a river by swinging from a vine. The vine is 10.0 m long, and his speed at the bottom of the swing (as he just clears the water) is 8.00 m/s. Tarzan doesn't know that the vine has a breaking strength of 1 000 N. Does he make it safely across the river?

FIG. P5.21

Solution   **Conceptualize.** The jungle lord does not move all the way around a circle, but he moves on an arc of a circle of radius 10 m. His direction of motion is changing, so he has change-in-direction acceleration, called centripetal acceleration. As he passes the bottom of the swing his speed is no longer increasing and not yet decreasing. He has no change-in-speed (tangential) acceleration but only acceleration $\dfrac{v^2}{r}$ upward.

Categorize. We can think of this as a straightforward particle-under-a-net-force problem. We will find the value of tension required under the assumption that Tarzan passes the bottom of the swing. If it is less than 1 000 N, he will make it across the river.

Analyze. The forces acting on Tarzan are the force of gravity $m\vec{g}$ and the force from the rope, $\vec{T}$. At the lowest point in his motion, $\vec{T}$ is upward and $m\vec{g}$ is downward as in the free-body diagram. Thus, Newton's second law gives

$$T - mg = \dfrac{mv^2}{r}$$

Solving for T, with $v = 8.00$ m/s, $r = 10.0$ m, and $m = 85.0$ kg, gives

continued on next page

$$T = m\left(g + \frac{v^2}{r}\right) = (85.0 \text{ kg})\left(9.80 \text{ m/s}^2 + \frac{(8.00 \text{ m/s})^2}{10.0 \text{ m}}\right) = 1.38 \times 10^3 \text{ N}$$

Since T **exceeds** the breaking strength of the vine (1 000 N), Tarzan **doesn't make it!** The vine breaks **before** he reaches the bottom of the swing.   ◊

Finalize. Tarzan's forward motion at the bottom is explained by Newton's first law. All of the forces on him are perpendicular to his motion. They explain his acceleration according to Newton's second law, as opposed to explaining his velocity.

P5.23   A pail of water is rotated in a vertical circle of radius 1.00 m. What is the minimum speed of the pail when it is upside down at the top of the circle if no water is to spill out?

Solution   The normal force, $\vec{n}$, will maintain exactly enough force to prevent the water from going **through** the bottom of the bucket. If water were to spill out, the force of gravity would exceed that required to provide the centripetal acceleration, and the normal force would be zero:

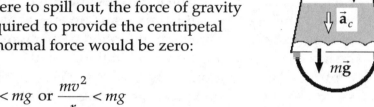

$$ma_c < mg \text{ or } \frac{mv^2}{r} < mg$$

FIG. P5.23

Since that is the only case in which water would spill out, all other cases must result in no water spilling out. That is,

$$\frac{mv^2}{r} \geq mg \text{ or } v^2 \geq rg$$

At the minimum speed, we have $v_{\text{min}} = \sqrt{rg} = \sqrt{(1.00 \text{ m})(9.80 \text{ m/s}^2)} = 3.13 \text{ m/s}$ ◊

P5.27   A small spherical bead of mass 3.00 g is released from rest at $t = 0$ in a bottle of liquid shampoo. The terminal speed is observed to be $v_T = 2.00 \text{ cm/s}$.

(a)   Find the value of the constant b in Equation 5.4.

(b)   Find the time τ at which the bead reaches $0.632 v_T$.

continued on next page

(c)   Find the value of the resistive force when the bead reaches terminal speed.

Solution (a)   The speed v varies with time according to Equation 5.6

$$v = \frac{mg}{b}\left(1 - e^{-bt/m}\right) = v_T\left(1 - e^{-bt/m}\right)$$

where $v_T = \dfrac{mg}{b}$ is the terminal speed. Hence,

$$b = \frac{mg}{v_T} = \frac{\left(3.00\times10^{-3}\ \text{kg}\right)\left(9.80\ \text{m/s}^2\right)}{2.00\times10^{-2}\ \text{m/s}} = 1.47\ \text{N}\cdot\text{s/m} \qquad \lozenge$$

(b)   To find the time for v to reach $0.632v_T$, we substitute $v = 0.632v_T$ into Equation 5.6, giving

$$0.632v_T = v_T\left(1 - e^{-bt/m}\right) \text{ or } 0.368 = e^{-(1.47t/0.003\ 00)}.$$

Solve for t by taking the natural logarithm of each side of the equation:

$$\ln(0.368) = -\frac{1.47t}{3.00\times10^{-3}} \text{ or } t = 2.04\times10^{-3}\ \text{s}. \qquad \lozenge$$

(c)   At terminal speed, $R = v_T b = mg$.

Therefore, $R = \left(3.00\times10^{-3}\ \text{kg}\right)\left(9.80\ \text{m/s}^2\right) = 2.94\times10^{-2}\ \text{N}.$ \qquad $\lozenge$

P5.29   A motor boat cuts its engine when its speed is 10.0 m/s and coasts to rest. The equation describing the motion of the motorboat during this period is $v = v_i e^{-ct}$, where v is the speed at time t, v_i is the initial speed, and c is a constant. At $t = 20.0$ s, the speed is 5.00 m/s.

(a)   Find the constant c.

(b)   What is the speed at $t = 40.0$ s?

(c)   Differentiate the expression for $v(t)$ and thus show that the acceleration of the boat is proportional to the speed at any time.

continued on next page

Solution (a)   We must fit the equation $v = v_i e^{-ct}$ to the two data points:

At $t = 0$, $v = 10.0$ m/s: $v = v_i e^{-ct}$

$$10.0 \text{ m/s} = v_i e^0 = v_i \times 1$$

$$v_i = 10.0 \text{ m/s}$$

At $t = 20.0$ s, $v = 5.00$ m/s: $5.00 \text{ m/s} = (10.0 \text{ m/s})e^{-c(20.0 \text{ s})}$

$$0.500 = e^{-c(20.0 \text{ s})}$$

$$\ln(0.500) = (-c)(20.0 \text{ s})$$

$$c = \frac{-\ln(0.500)}{20.0 \text{ s}} = 0.034\,7 \text{ s}^{-1}$$ ◊

(b)   At all times $v = (10.0 \text{ m/s})e^{-(0.034\,7)t}$

At $t = 40.0$ s, $v = (10.0 \text{ m/s})e^{-(0.034\,7)(40.0)} = 2.50 \text{ m/s}$ ◊

(c)   The acceleration is the rate of change of the velocity:

$$a = \frac{dv}{dt} = \frac{d}{dt}v_i e^{-ct} = v_i\left(e^{-ct}\right)(-c) = -c\left(v_i e^{-ct}\right) = -cv = \left(-0.0347 \text{ s}^{-1}\right)v$$

Thus, the acceleration is a negative constant times the speed.   ◊

P5.33   When a falling meteor is at a distance above the Earth's surface of 3.00 times the Earth's radius, what is its free-fall acceleration caused by the gravitational force exerted on it?

Solution   The acceleration of gravity due to the Earth, $g = \dfrac{GM_E}{r^2}$, follows an inverse-square law. At the surface (at distance one Earth-radius R_E from the center), it is 9.80 m/s^2.

At an altitude $3.00R_E$ above the surface (at distance $4.00R_E$ from the center), the acceleration of gravity will be $4.00^2 = 16.0$ times smaller:

$$g = \frac{GM_E}{(4.00R_E)^2} = \frac{GM_E}{16.0R_E{}^2} = \frac{9.80 \text{ m/s}^2}{16.0} = 0.612 \text{ m/s}^2 \text{ down}$$ ◊

P5.49   Because the Earth rotates about its axis, a point on the equator experiences a centripetal acceleration of 0.033 7 m/s^2, whereas a point at the poles experiences no centripetal acceleration.

(a)   Show that at the equator the gravitational force on an object must exceed the normal force required to the support the object. That is, show that the object's true weight exceeds its apparent weight.

(b)   What is the apparent weight at the Equator and at the poles of a person having a mass of 75.0 kg? (Assume that the Earth is a uniform sphere and take $g = 9.800 \text{ m/s}^2$.)

Solution   **Conceptualize:** Since the centripetal acceleration is a small fraction (~0.3%) of g, we should expect that a person would have an apparent weight is just slightly less at the equator than at the poles due to the rotation of the Earth.

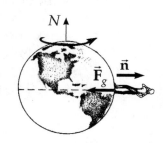

Categorize: We will apply Newton's second law and the equation for centripetal acceleration.

Analyze: **FIG. P5.49**

(a)   Let $\bar{\mathbf{n}}$ represent the force exerted on the person by the scale, which is the "apparent weight." The true weight is $m\bar{\mathbf{g}}$. Summing up forces on the object in the direction towards the Earth's center gives

$$mg - n = ma_c \qquad (1)$$

where $a_c = \dfrac{v^2}{R_e} = 0.033\,7 \text{ m/s}^2$ is the centripetal acceleration directed toward the center of the Earth. Thus, we see that $n = m(g - a_c) < mg$

$$mg = n + ma_c > n \qquad (2) \qquad \lozenge$$

(b)   If $m = 75.0$ kg and $g = 9.800 \text{ m/s}^2$, at the Equator:

$$n = m(g - a_c) = (75.0 \text{ kg})(9.800 \text{ m/s}^2 - 0.033\,7 \text{ m/s}^2)$$
$$n = 732 \text{ N} \qquad\qquad \lozenge$$

at the Poles: $a_c = 0$: $n = mg = (75.0 \text{ kg})(9.80 \text{ m/s}^2) = 735 \text{ N}$ $\qquad \lozenge$

continued on next page

Finalize: As we expected, the person does appear to weigh about 0.3% less at the equator than the poles. We might extend this problem to consider the effect of the earth's bulge on a person's weight. Since the earth is fatter at the equator than the poles, would you expect g to be less than 9.80 m/s^2 at the equator and slightly more at the poles?

P5.55 An amusement park ride consists of a large vertical cylinder that spins about its axis fast enough that any person inside is held up against the wall when the floor drops away (Fig. P5.55). The coefficient of static friction between a person and the wall is μ_s, and the radius of the cylinder is R.

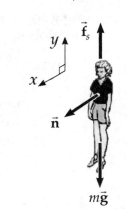

(a) Show that the maximum period of revolution necessary to keep the person from falling is given by $T = \left(\dfrac{4\pi^2 R \mu_s}{g} \right)^{1/2}$.

(b) Obtain a numerical value for T assuming that $R = 4.00$ m and $\mu_s = 0.400$. How many revolutions per minute does the cylinder make?

Solution We model the person as a particle in uniform circular motion.

(a) The wall's normal force pushes inward:

$$n = \frac{mv^2}{R} = \frac{m}{R}\left(\frac{2\pi R}{T}\right)^2 = \frac{4\pi^2 Rm}{T^2}$$

FIG. P5.55

The friction and weight balance: $f_s = \mu_s n = mg$. Therefore, with

$\mu_s n = mg$, $\mu_s n = \mu_s \dfrac{4\pi^2 Rm}{T^2} = mg$. Solving, $T^2 = \dfrac{4\pi^2 R \mu_s}{g}$ gives

$T = \sqrt{\dfrac{4\pi^2 R \mu_s}{g}}$. The person's mass has divided out. An employee of the park does not have to stand at the entrance turning away especially heavy riders. On the other hand, a rider wearing a plastic raincoat, with small μ_s, would slide down the wall unless the rotation rates were increased sufficiently. Mathematically, T is proportional to the square root of μ_s.

continued on next page

(b) $T = \sqrt{\dfrac{4\pi^2(4.00\text{ m})(0.400)}{9.80\text{ m/s}^2}} = 2.54\text{ s}$

The angular speed is $\left(\dfrac{1\text{ rev}}{2.54\text{ s}}\right)\left(\dfrac{60\text{ s}}{\text{min}}\right) = 23.6\text{ rev/min}$ ◊

Related Questions:

(a)   Why is the normal force horizontally inward?

(b)   Why is there no upward normal force?

(c)   Why is the frictional force directed upward?

(d)   Why is it not kinetic friction?

(e)   Why is there no outward force on her?

Answers:

(a)   Because the wall is vertical, and on the outside.

(b)   Because there is no floor.

(c)   The friction opposes the possible relative motion of the person sliding down the wall.

(d)   Because person and wall are moving together, stationary with respect to each other.

(e)   No other object pushes out on her. She pushes out on the wall as the wall pushes inward on her.

P5.59 A model airplane of mass 0.750 kg flies in a horizontal circle at the end of a 60.0-m control wire, with a speed of 35.0 m/s. Compute the tension in the wire assuming that it makes a constant angle of 20.0° with the horizontal. The forces exerted on the airplane are the pull of the control wire, the gravitational force, and aerodynamic lift, which acts at 20.0° inward from the vertical as shown in Figure P5.59.

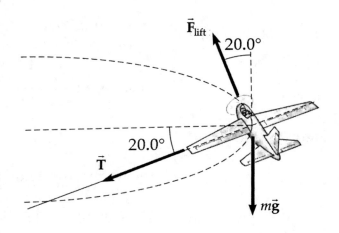

FIG. P5.59

Solution The plane's acceleration is toward the center of the circle of motion, so it is horizontal. The radius of the circle of motion is (60.0 m) cos 20.0° = 56.4 m, and the acceleration is

$$a_c = \frac{v^2}{r} = \frac{(35.0 \text{ m/s})^2}{56.4 \text{ m}} = 21.7 \text{ m/s}^2$$

We can also calculate the weight of the airplane:

$$F_g = mg = (0.750 \text{ kg})(9.80 \text{ m/s}^2) = 7.35 \text{ N}.$$

We define our axes for convenience. In this case, two of the forces—one of them our force of interest—are directed along the 20.0° lines.

We define the x-axis to be directed in the $(+\vec{T})$ direction, and the y-axis to be directed in the direction of lift. With these definitions, the x component of the centripetal acceleration is

$$a_{cx} = a_c \cos(20.0°)$$

and $\sum F_x = ma_x$ yields $T + F_g \sin(20.0°) = ma_{cx}$. Solving for T,
$T = ma_{cx} - F_g \sin(20.0°)$,

$$T = (0.750 \text{ kg})(21.7 \text{ m/s}^2)(\cos 20.0°) - (7.35 \text{ N})\sin 20.0°$$

and $T = (15.3 \text{ N}) - (2.51 \text{ N}) = 12.8 \text{ N}$.   ◊

Energy and Energy Transfer

Section 6.2   Work Done By a Constant Force

Work done by a constant force is defined as the product of the component of the force in the direction of the displacement and the magnitude of the displacement. The **unit of work** in the SI system is the newton·meter, N·m: 1 newton·meter = 1 joule (J).

Section 6.3   The Scalar Product of Two Vectors

The scalar product or dot product of any two vectors is a scalar quantity equal to the product of the magnitudes of the two vectors and the cosine of the angle included between the directions of the two vectors.

Section 6.4   Work Done By a Varying Force

Work done by a varying force is equal to the area under the force-displacement curve.

Section 6.5   Kinetic Energy and the Work-Kinetic Energy Theorem

The work done by a net or resultant force in displacing a particle equals the change in the kinetic energy of the particle. **This is known as the work-kinetic energy theorem**.

Section 6.8   Power

Power is the time rate of doing work or expending energy. The SI unit of power is the watt, W; $1\ W = 1\ J/s$.

EQUATIONS AND CONCEPTS

The **work done on a system by an a constant force** is defined to be the product of the displacement and the component of force in the direction of the displacement. *Work is a scalar quantity.*

$$W \equiv F\Delta r \cos\theta \tag{6.1}$$

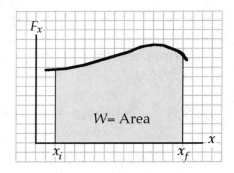

The **scalar (or dot) product** of any two vectors $\vec{A}$ and $\vec{B}$ is defined to be a scalar quantity whose magnitude is $AB\cos\theta$. (A and B are the magnitudes of the two vectors and θ is the smaller of the two angles between the directions of $\vec{A}$ and $\vec{B}$.)

$$\vec{A} \cdot \vec{B} \equiv AB\cos\theta \tag{6.3}$$

The scalar product can be expressed in terms of the x, y, and z components of the two vectors.

$$\vec{A} \cdot \vec{B} = A_x B_x + A_y B_y + A_z B_z \tag{6.9}$$

The **work done by a constant force** can be expressed as the dot product (scalar product) of the force and displacement vectors.

$$W = \vec{F} \cdot \Delta\vec{r} = F\Delta r \cos\theta \tag{6.4}$$

The **work done by a variable force** on a particle that has been displaced along the x axis from x_i to x_f is given by an integral expression.

$$W = \int_{x_i}^{x_f} F_x \, dx \tag{6.11}$$

Graphically, the work done equals the area under the F_x versus x curve.

The **work done by the spring force** $(-kx)$ as the "free end" of the spring undergoes an arbitrary displacement from x_i to x_f is given by Equation 6.15.

$$W_s = \int_{x_i}^{x_f} (-kx)\,d\dot{x} = \frac{1}{2}kx_i^2 - \frac{1}{2}kx_f^2 \tag{6.15}$$

Note that work done by a force can be positive, negative, or zero depending on the value of θ. Work done by a force is positive if $\vec{F}$ has a component in the direction of $\Delta\vec{r}(0 \le \theta < 90°)$; W is negative if the projection of $\vec{F}$ onto $\Delta\vec{r}$ is opposite to $\Delta\vec{r}(90° < \theta < 180°)$. Finally, W is zero if $\vec{F}$ is perpendicular to $\Delta\vec{r}(\theta = \pm90°)$. *Work is a scalar quantity, which has SI units of joules (J), where* $1\,J = 1\,N \cdot m$.

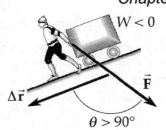

In the figure above, the work done by force $\vec{F}$ as the object is displaced by $\Delta\vec{r}$ is negative.

Kinetic energy, K, is energy associated with the motion of an object. *Kinetic energy is a scalar quantity and has the same units as work.*

$$K \equiv \frac{1}{2}mv^2 \qquad (6.18)$$

The **work-kinetic energy theorem** states that the net work done on a body equals the change in kinetic energy of the body. The speed of a body will only change if there is net work done on it. *The work-kinetic energy theorem is valid for a particle or for a system that can be modeled as a particle.*

$$W_{net} = K_f - K_i = \Delta K \qquad (6.19)$$

$$W_{net} = \frac{1}{2}mv_f^2 - \frac{1}{2}mv_i^2 \qquad (6.17)$$

The **principle of conservation of energy** is described by Equation 6.20. E represents the total energy of a system and T is the quantity of energy transferred across the system boundary by any transfer mechanism.

$$\Delta E_{system} = \sum T \qquad (6.20)$$

The work-kinetic energy equation takes the form of Equation 6.23 when friction is included among the forces acting on an object.

$$\sum W_{other\ forces} - f_k d = \Delta K \qquad (6.23)$$

The **internal energy of a system** will change as a result of transfer of energy across the system boundary by frictional forces. *The increase in internal energy is equal to the decrease in kinetic energy.*

$$\Delta E_{int} = f_k d \qquad (6.24)$$

The **average power** supplied by a force is the ratio of the work done by that force to the time interval over which it acts.

$$\mathscr{P}_{ave} \equiv \frac{W}{\Delta t} \qquad (6.25)$$

The **instantaneous power** is equal to the limit of the average power as the time interval approaches zero.

$$\mathscr{P} = \frac{dW}{dt} = \vec{\mathbf{F}} \cdot \frac{d\vec{\mathbf{r}}}{dt} = \vec{\mathbf{F}} \cdot \vec{\mathbf{v}} \tag{6.27}$$

The **SI unit of power** is J/s, which is called a watt (W). The kilowatt-hour is a unit of energy. The unit of power in the U.S. Customary system is the horsepower.

$$1\ \text{W} = 1\ \text{J/s} = 1\ \text{kg} \cdot \text{m}^2/\text{s}^2$$

$$1\ \text{kWh} = 3.60 \times 10^6\ \text{J}$$

$$1\ \text{hp} = 746\ \text{W}$$

SUGGESTIONS, SKILLS, AND STRATEGIES

There are two new mathematical skills you must learn. The first is the definition of the scalar (or dot) product, $\vec{\mathbf{A}} \cdot \vec{\mathbf{B}} \equiv AB\cos\theta$ where θ is the angle between $\vec{\mathbf{A}}$ and $\vec{\mathbf{B}}$. Since $\vec{\mathbf{A}} \cdot \vec{\mathbf{B}}$ is a scalar, the order of the terms can be interchanged. That is, $\vec{\mathbf{A}} \cdot \vec{\mathbf{B}} = \vec{\mathbf{B}} \cdot \vec{\mathbf{A}}$. Furthermore, $\vec{\mathbf{A}} \cdot \vec{\mathbf{B}}$ can be positive, negative, or zero depending on the value of θ. (That is, $\cos\theta$ varies from -1 to $+1$.) If vectors are expressed in unit vector form, then the dot product is conveniently carried out using the multiplication table for unit vectors:

$$\hat{\mathbf{i}} \cdot \hat{\mathbf{i}} = \hat{\mathbf{j}} \cdot \hat{\mathbf{j}} = \hat{\mathbf{k}} \cdot \hat{\mathbf{k}} = 1; \qquad \hat{\mathbf{i}} \cdot \hat{\mathbf{j}} = \hat{\mathbf{i}} \cdot \hat{\mathbf{k}} = \hat{\mathbf{j}} \cdot \hat{\mathbf{k}} = 0$$

The second operation introduced in this chapter is the definite integral. In Section 6.3, it is shown that the work done by a variable force F_x in displacing a particle a small distance Δx is given by

$$\Delta W \approx F_x \Delta x$$

(ΔW equals the area of the shaded rectangle in Figure 6.1.) The total work done by F_x as the particle is displaced from x_i to x_f is given approximately by the sum of such terms. If we take such a sum, letting the widths of the displacements approach dx, the number of terms in the sum becomes very large and we get the actual work done:

$$W = \lim_{\Delta x \to 0} \sum_{x_i}^{x_f} F_x \Delta x = \int_{x_i}^{x_f} F_x \, dx$$

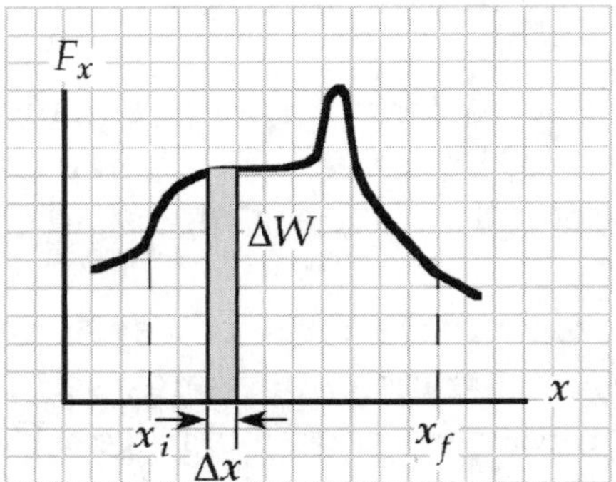

FIG. 6.1

The quantity on the right is a definite integral, which graphically represents the area under the F_x versus x curve, as in Figure 6.1.

REVIEW CHECKLIST

✓ Define the work done by a constant force, and realize that work is a scalar. Describe the work done by a force that varies with position. In the one-dimensional case, note that the work done equals the area under the F_x versus x curve.

✓ Take the scalar or dot product of any two vectors $\vec{A}$ and $\vec{B}$ using the definition $\vec{A} \cdot \vec{B} \equiv AB \cos\theta$, or by writing $\vec{A}$ and $\vec{B}$ in unit vector form and using the multiplication table for unit vectors.

✓ Define the kinetic energy of an object of mass m moving with a speed v.

✓ Relate the work done by the net force on an object to the change in kinetic energy. The relation $W_{net} = \Delta K = K_f - K_i$ is called the work-kinetic energy theorem, and is valid whether or not the (resultant) force is constant. That is, if we know the net work done on a particle as it undergoes a displacement, we also know the change in its kinetic energy. This is the most important concept in this chapter, so you must understand it thoroughly.

✓ Define the concepts of average power and instantaneous power (the time rate of doing work).

ANSWERS TO SELECTED QUESTIONS

Q6.5 As a simple pendulum swings back and forth, the forces acting on the suspended mass are the gravitational force, the tension in the supporting cord, and air resistance.

    (a)   Which of these forces, if any, does no work on the pendulum?

    (b)   Which of these forces does negative work at all times during its motion?

    (c)   Describe the work done by the gravitational force while the pendulum is swinging.

Answer   (a)   The tension in the supporting cord does no work, because the motion of the pendulum is always perpendicular to the cord, and therefore to the force exerted by the string.

continued on next page

(b)   The air resistance does negative work at all times, since the air resistance is always acting in a direction opposite to the motion.

(c)   The weight always acts downward; therefore, the work done by the gravitational force is positive on the downswing, and negative on the upswing.

Q6.9   Can kinetic energy be negative? Explain.

Answer   No. Kinetic energy $= \dfrac{mv^2}{2}$. Since v^2 is always positive, K is always positive.

Q6.11   One bullet has twice the mass of a second bullet. If both are fired so that they have the same speed, which has more kinetic energy? What is the ratio of the kinetic energies of the two bullets?

Answer   The kinetic energy of the more massive bullet is twice that of the lower mass bullet.

Q6.13   (a)   If the speed of a particle is doubled, what happens to its kinetic energy?

(b)   What can be said about the speed of a particle if the net work done on it is zero?

Answer   (a)   Kinetic energy, K, depends on the square of the velocity. Therefore if the speed is doubled, the kinetic energy will increase by a factor of four.

(b)   If net work is zero, the speed of a particle does not change.

SOLUTIONS TO SELECTED PROBLEMS

P6.1   A block of mass 2.50 kg is pushed 2.20 m along a frictionless horizontal table by a constant 16.0-N force directed 25.0° below the horizontal. Determine the work done on the block by

(a)   the applied force.

(b)   the normal force exerted by the table.

(c)   the gravitational forced.

(d)   Determine the total work done on the block.

continued on next page

Solution $W = F\Delta r \cos\theta$

(a) By the applied force, $W_{app} = (16.0\ \text{N})(2.20\ \text{m})\cos(25.0°) = 31.9\ \text{J}$. ◊

(b) By normal force, $W_n = n\Delta r \cos\theta = n\Delta r \cos(90°) = 0$. ◊

(c) By force of gravity, $W_g = F_g\Delta r \cos\theta = mg\Delta r \cos(90°) = 0$. ◊

(d) Net work done on the block: $W_{net} = W_{app} + W_n + W_g = 31.9\ \text{J}$. ◊

P6.3 Batman, whose mass is 80.0 kg, is dangling on the free end of a 12.0-m rope, the other end of which is fixed to a tree limb above. He is able to get the rope in motion as only Batman knows how, eventually getting it to swing enough that he can reach a ledge when the rope makes a 60.0° angle with the vertical. How much work was done by the gravitational force on the superhero in this maneuver?

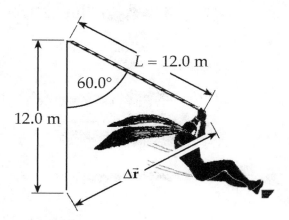

FIG. P6.3

Solution The work done is $W = \vec{\mathbf{F}} \cdot \Delta\vec{\mathbf{r}}$, where the gravitational force is

$$\vec{\mathbf{F}}_g = -mg\,\hat{\mathbf{j}} = (80.0\ \text{kg})\left(-9.80\ \text{m/s}^2\ \hat{\mathbf{j}}\right) = -784\,\hat{\mathbf{j}}\ \text{N}.$$

The superhero travels

$$\Delta\vec{\mathbf{r}} = L\sin60°\,\hat{\mathbf{i}} + L(1 - \cos60°)\,\hat{\mathbf{j}}$$
$$= (12.0\ \text{m})\sin60°\,\hat{\mathbf{i}} + (12.0\ \text{m})(1 - \cos60°)\,\hat{\mathbf{j}}$$

Thus,

$$W = \vec{\mathbf{F}} \cdot \Delta\vec{\mathbf{r}} = \left(-784\,\hat{\mathbf{j}}\ \text{N}\right) \cdot \left(10.4\,\hat{\mathbf{i}}\ \text{m} + 6.00\,\hat{\mathbf{j}}\ \text{m}\right) = -4.70 \times 10^3\ \text{J}.$$ ◊

P6.7   A force $\vec{F} = \left(6\hat{i} - 2\hat{j}\right)$ N acts on a particle that undergoes a displacement $\Delta\vec{r} = \left(3\hat{i} + \hat{j}\right)$ m.

(a)   Find the work done by the force on the particle.

(b)   Find the angle between $\vec{F}$ and $\Delta\vec{r}$.

Solution   We use the mathematical representation of the definition of work.

(a) $W = \vec{F} \cdot \Delta\vec{r} = \left(6\hat{i} - 2\hat{j}\right)$ N $\cdot \left(3\hat{i} + 1\hat{j}\right)$ m $= (6$ N$)(3$ m$) + (-2$ N$)(1$ m$) = 18$ J $- 2$ J

   $W = 16.0$ J   ◊

(b) $\left|\vec{F}\right| = \sqrt{F_x^2 + F_y^2} = \sqrt{6^2 + (-2)^2}$ N $= 6.32$ N

   $\left|\Delta\vec{r}\right| = \sqrt{\Delta r_x^2 + \Delta r_y^2} = \sqrt{3^2 + 1^2}$ m $= 3.16$ m

   $W = F\Delta r \cos\theta$: $\cos\theta = \dfrac{W}{F\Delta r} = \dfrac{16.0 \text{ J}}{(6.32 \text{ N})(3.16 \text{ m})} = 0.800$

   and $\theta = \cos^{-1}(0.800) = 36.9°$.   ◊

P6.11   A particle is subject to a force F_x that varies with position as in Figure P6.11. Find the work done by the force on the object as it moves

(a)   from $x = 0$ to $x = 5.00$ m,

(b)   from $x = 5.00$ m to $x = 10.0$ m

(c)   from $x = 10.0$ m to $x = 15.0$ m.

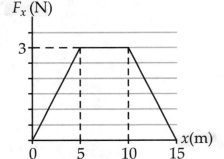

FIG. P6.11

(d)   What is the total work done by the force over the distance $x = 0$ to $x = 15.0$ m?

Solution $W = \int F_x dx$. We use the graphical representation of the definition of work. W equals the area under the force-displacement curve.

continued on next page

(a)   For the region $0 \le x \le 5.00$ m, $W = \dfrac{(3.00 \text{ N})(5.00 \text{ m})}{2} = 7.50$ J.   ◊

(b)   For the region 5.00 m $\le x \le 10.0$ m, $W = (3.00 \text{ N})(5.00 \text{ m}) = 15.0$ J.   ◊

(c)   For the region 10.0 m $\le x \le 15.0$ m, $W = \dfrac{(3.00 \text{ N})(5.00 \text{ m})}{2} = 7.50$ J.   ◊

(d)   For the region $0 \le x \le 15.0$ m, $W = (7.50 + 7.50 + 15.0)$ J $= 30.0$ J.   ◊

P6.17   It takes 4.00 J of work to stretch a Hooke's-law spring 10.0 cm from its unstressed length. Determine the extra work required to stretch it an additional 10.0 cm.

Solution **Conceptualize:** We know that the force required to stretch a spring is proportional to the distance the spring is stretched, and since the work required is proportional to the force **and** to the distance, then $W \propto x^2$. This means if the extension of the spring is doubled, the work will increase by a factor of 4, so that for $x = 20$ cm, $W = 16$ J, requiring 12 J of additional work.

Categorize: Let's confirm our answer using Hooke's law and the definition of work.

Analyze: The linear spring force relation is given by Hooke's law: $F_s = -kx$. Integrating with respect to x, we find the work done by the spring is:

$$W_s = \int\limits_{x_i}^{x_f} F dx = \int\limits_{x_i}^{x_f} (-kx) dx = -\frac{1}{2}k\left(x_f^2 - x_i^2\right)$$

However, we want the work done **on** the spring, which is $W = -W_s = \dfrac{1}{2}k\left(x_f^2 - x_i^2\right)$. We know the work for the first 10 cm, so we can find the force constant:

$$k = \frac{2W}{x_f^2 - x_i^2} = \frac{2(4.00 \text{ J})}{(0.100 \text{ m})^2 - 0} = 800 \text{ N/m}$$

Substituting for k, x_i and x_f, the extra work for the next step of extension is

$$W = \frac{1}{2}(800 \text{ N/m})\left[(0.200 \text{ m})^2 - (0.100 \text{ m})^2\right] = 12.0 \text{ J}$$ ◊

Finalize: Our calculated answer agrees with our prediction. It is helpful to remember that the force required to stretch a spring is proportional to the distance the spring is extended, but the work is proportional to the square of the extension.

P6.25   A 2 100-kg pile driver is used to drive a steel I-beam into the ground. The pile driver falls 5.00 m before coming into contact with the top of the beam, and it drives the beam 12.0 cm farther into the ground before coming to rest. Using energy considerations, calculate the average force the beam exerts on the pile driver while the pile driver is brought to rest.

Solution   **Conceptualize:** Anyone who has hit their thumb with a hammer knows that the resulting force is greater than just the weight of the hammer, so we should also expect the force of the pile driver to be significantly greater than its weight: $F \gg mg \approx 20$ kN. The force **on** the pile driver will be directed upwards.

Categorize: The average force stopping the driver can be found from the work that results from the gravitational force starting its motion. The initial and final kinetic energies are zero.

Analyze: Choose the initial point when the mass is elevated and the final point when it comes to rest again 5.12 m below. Two forces do work on the pile driver: gravity (weight) and the normal force exerted by the beam on the pile driver. $K_i + \sum W = K_f$: so that $0 + mgd_w \cos(0) + nd_n \cos(180°) = 0$ where $d_w = 5.12$ m, $d_n = 0.120$ m, and $m = 2\,100$ kg. In this situation, the weight vector is in the direction of motion and the beam exerts a force on the pile driver that is opposite the direction of motion.

$$(2\,100 \text{ kg})(9.80 \text{ m/s}^2)(5.12 \text{ m}) + n(0.120 \text{ m})(-1) = 0$$

Solve for n: $n = \dfrac{1.05 \times 10^5 \text{ J}}{0.120 \text{ m}} = 878$ kN.   ◊

Finalize: The normal force is larger than 20 kN as we expected, and is actually about 43 times greater than the weight of the pile driver, which is why this machine is so effective.

Additional Calculation: Show that the work done by gravity on an object can be represented by mgh, where h is the vertical height that the object falls. Apply your results to the problem above.

   By the figure to the right, where $\Delta \vec{r}$ is the path of the object, and h is the height that the object falls,

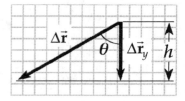

FIG. P6.25

$$h = \left| \Delta \vec{r}_y \right| = \Delta r \cos \theta.$$

continued on next page

Since $F = mg$, $mgh = F\Delta r \cos\theta = \vec{F} \cdot \Delta\vec{r}$. ◊

In this problem, $mgh = n(d_n)$ so $n = \dfrac{mgh}{d_n}$

and $n = \dfrac{(2\,100 \text{ kg})(9.80 \text{ m/s}^2)(5.12 \text{ m})}{(0.120 \text{ m})} = 878 \text{ kN}$. ◊

P6.29   A 40.0-kg box initially at rest is pushed 5.00 m along a rough, horizontal floor with a constant applied horizontal force of 130 N. The coefficient of the friction between box and floor is 0.300. Find

(a)   the work done by the applied force.

(b)   the increase in internal energy in the box-floor system as a result of friction.

(c)   the work done by the normal force.

(d)   the work done by the gravitational force.

(e)   the change in kinetic energy of the box.

(f)   the final speed of the box.

Solution $\mu_k = 0.300$

$\Delta r = 5.00$ m

$v_i = 0$

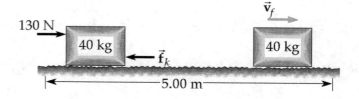

FIG. P6.29

(a)   The applied force and the motion are both **horizontal**.

$$W_F = \vec{F} \cdot \Delta\vec{r} = F\Delta r \cos(0°) = (130 \text{ N})(5.00 \text{ m})(1) = 650 \text{ J}$$ ◊

(b) $f_k = \mu_k n = \mu_k mg = 0.300(40.0 \text{ kg})(9.80 \text{ m/s}^2) = 117.6 \text{ N}$

$\Delta E_{int} = f_k d = (117.6 \text{ N})(5.00 \text{ m}) = 588 \text{ J}$ ◊

(c)   Since the normal force is perpendicular to the motion,

$$W_n = F\Delta r \cos(90°) = (392 \text{ N})(5.00 \text{ m})(0) = 0.$$ ◊

continued on next page

(d)   The force of gravity is also perpendicular to the motion, so $W_g = 0$. ◊

We use the energy version of the nonisolated system model.

(e) $\Delta K = W_{\text{other forces}} - f_k d = 650 \text{ J} - 588 \text{ J} = 62.0 \text{ J}$ ◊

(f) $\frac{1}{2} m v_i^2 + W_{\text{other forces}} - f_k d = \frac{1}{2} m v_f^2$

$v_f = \sqrt{\frac{2}{m}\left[\Delta K + \frac{1}{2} m v_i^2\right]} = \sqrt{\left(\frac{2}{40.0 \text{ kg}}\right)\left[62.0 \text{ J} + \frac{1}{2}(40.0 \text{ kg})(0)^2\right]} = 1.76 \text{ m/s}$ ◊

P6.31   A crate of mass 10.0 kg is pulled up a rough incline with an initial speed of 1.50 m/s. The pulling force is 100 N parallel to the incline, which makes an angle of 20.0° with the horizontal. The coefficient of kinetic friction is 0.400, and the crate is pulled 5.00 m.

(a)   How much work is done by the gravitational force on the crate?

(b)   Determine the increase in internal energy of the crate-incline system owing to friction.

(c)   How much work is done by the 100-N force on the crate?

(d)   What is the change in kinetic energy of the crate?

(e)   What is the speed of the crate after being pulled 5.00 m?

Solution The force of gravity is $(10.0 \text{ kg})(9.80 \text{ m/s}^2) = 98.0 \text{ N}$ straight down, at an angle of $(90.0° + 20.0°) = 110.0°$ with the motion. The work done by gravity on the crate is

(a) $W_g = \vec{F} \cdot \Delta \vec{r} = (98.0 \text{ N})(5.00 \text{ m})\cos 110.0° = -168 \text{ J}$ ◊

(b)   Setting the xy axes parallel and perpendicular to the incline,

from $\Sigma F_y = m a_y$, $+n - (98.0 \text{ N})\cos 20.0° = 0$,

$n = 92.1 \text{ N}$

and $f_k = \mu_k n = 0.400(92.1 \text{ N}) = 36.8 \text{ N}$.

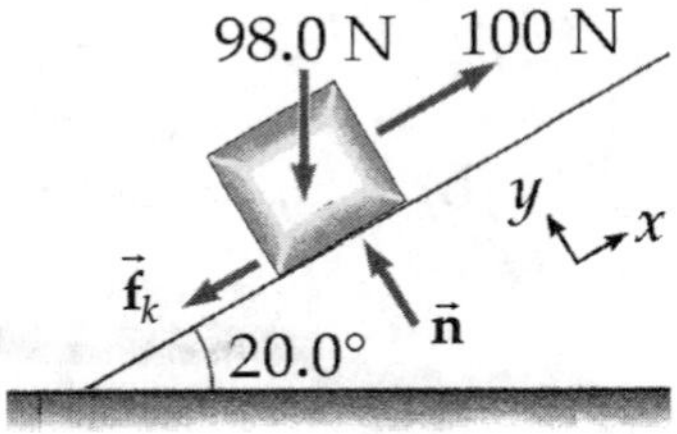

FIG. P6.31

continued on next page

Therefore, $\Delta E_{\text{int}} = f_k d = (36.8 \text{ N})(5.00 \text{ m}) = 184 \text{ J}$.   ◊

(c) $W = \vec{\mathbf{F}} \cdot \Delta \vec{\mathbf{r}} = 100 \text{ N}(5.00 \text{ m})\cos(0°) = +500 \text{ J}$ ◊

(d)   We use the energy version of the nonisolated system model.

$$\Delta K = -f_k d + \sum W_{\text{other forces}}$$

$$\Delta K = -f_k d + W_g + W_{\text{applied force}} + W_n$$

$$\Delta K = -184 \text{ J} - 168 \text{ J} + 500 \text{ J} + 0 = 148 \text{ J}$$ ◊

The normal force does zero work, because it is at 90° to the motion.

(e)   Since $K_f = K_i + \Delta K$, $\dfrac{1}{2}(10.0 \text{ kg}) v_f{}^2 = \dfrac{1}{2}(10.0 \text{ kg})(1.50 \text{ m/s})^2 + 148 \text{ J} = 159 \text{ J}$.

Thus, $v_f = \sqrt{\dfrac{2(159 \text{ kg} \cdot \text{m}^2/\text{s}^2)}{10.0 \text{ kg}}} = 5.65 \text{ m/s}$.   ◊

P6.33   A sled of mass m is given a kick on a frozen pond. The kick imparts to it an initial speed of 2.00 m/s. The coefficient of kinetic friction between sled and ice is 0.100. Use energy considerations to find the distance the sled moves before it stops.

Solution   **Conceptualize:** Since the sled's initial speed of 2 m/s (~4 mph) is reasonable for a moderate kick, we might expect the sled to travel several meters before coming to rest.

Categorize: We could solve this problem using Newton's second law, but we are asked to use the work-kinetic energy theorem: $W - f_k d = \Delta K = K_f - K_i$, where the change in kinetic energy of the sled results only from the friction between the sled and ice. (The weight and normal force both act at 90° to the motion, and therefore do no work on the sled.)

Analyze: The change in kinetic energy due to friction is $-f_k d$

where $f_k = \mu_k n = \mu_k m g$.

Since the final kinetic energy is zero, $W - f_k d = \Delta K = 0 - K_i = -\dfrac{1}{2} m v_i^2$.

So $-\mu_k m g d = -\dfrac{1}{2} m v_i^2$.

continued on next page

Thus, $d = \dfrac{mv_i^2}{2f_k} = \dfrac{mv_i^2}{2\mu_k mg} = \dfrac{v_i^2}{2\mu_k g} = \dfrac{(2.00 \text{ m/s})^2}{2(0.100)(9.80 \text{ m/s}^2)} = 2.04 \text{ m.}$ ◊

Finalize: The distance agrees with the prediction. It is interesting that the distance does not depend on the mass and is proportional to the square of the initial velocity. This means that a small car and a massive truck should be able to stop within the same distance if they both skid to a stop from the same initial speed. Also, doubling the speed requires four times as much stopping distance. Transportation safety officers often advise maintaining at least a 2-second gap between vehicles, as opposed to a fixed distance like 100 feet. This advice ensures a stopping region that increases linearly with speed. We invite you to formulate a simple rule that drivers can follow to get a cushion of space proportional to the square of speed.

P6.35   A 700-N Marine in basic training climbs a 10.0-m vertical rope at a constant speed in 8.00 s. What is his power output?

Solution   The marine must exert a 700 N upward force opposite the gravitational force to lift his body at constant speed. Then his muscles do work:

$$W = \vec{F} \cdot \Delta \vec{r} = \left(700\hat{j} \text{ N}\right) \cdot \left(10.0\hat{j} \text{ m}\right) = 7\,000 \text{ J.}$$

The power he puts out is $\mathscr{P}_{\text{avg}} = \dfrac{W}{\Delta t} = \dfrac{7\,000 \text{ J}}{8.00 \text{ s}} = 875 \text{ W.}$ ◊

P6.45   A 4.00-kg particle moves along the x axis. Its position varies with time according to $x = t + 2.0t^3$, where x is in meters and t is in seconds.

(a)   Find the kinetic energy at any time t.

(b)   Find the acceleration of the particle and the force acting on it at time t.

(c)   Find the power being delivered to the particle at time t.

(d)   Find the work done on the particle in the interval $t = 0$ to $t = 2.00$ s.

continued on next page

Solution   Given $m = 4.00$ kg and $x = t + 2.0t^3$, we find

(a) $v = \dfrac{dx}{dt} = \dfrac{d}{dt}\left(t + 2.0t^3\right) = 1.00 + 6.0t^2$

$K = \dfrac{1}{2}mv^2 = \dfrac{1}{2}(4.00 \text{ kg})\left(1.00 + 6.0t^2\right)^2 = \left(2.00 + 24t^2 + 72t^4\right) \text{J}$ ◊

(b) $a = \dfrac{dv}{dt} = \dfrac{d}{dt}\left(1.00 + 6.0t^2\right) = 12t \ \text{m}/\text{s}^2$ ◊

$F = ma = (4.00 \text{ kg})(12t) = 48t \text{ N}$ ◊

(c) $\mathscr{P} = \dfrac{dW}{dt} = \dfrac{dK}{dt} = \dfrac{d}{dt}\left(2.00 + 24t^2 + 72t^4\right) = \left(48t + 288t^3\right) \text{W}$ ◊

[or use $\mathscr{P} = Fv = 48t\left(1.00 + 6.0t^2\right)$]

(d) $W = K_f - K_i$ where $t_i = 0$ and $t_f = 2.00$ s.

At $t_i = 0$, $K_i = 2.00$ J

At $t_f = 2.00$ s, $K_i = \left[2.00 + 24(2.00 \text{ s})^2 + 72(2.00 \text{ s})^4\right] = 1\,250$ J

Therefore, $W = 1\,248 \text{ J} = 1.25 \times 10^3$ J   ◊

[or use $W = \displaystyle\int_{t_i}^{t_f} \mathscr{P} dt = \int_{0}^{2} \left(48t + 288t^3\right) dt$, etc.]

The given position function implies that the net force on the particle increases in direct proportion to the time. The particle does not move with constant acceleration but with greater and greater acceleration as time goes on. The power delivered to the particle contains a term proportional to time cubed, so it increases more steeply as time goes on. The kinetic energy expression goes as time to the fourth power for large values of time. If the expression continues to be an accurate model, the kinetic energy at 2 000 s will be about 16 times larger than the kinetic energy at 1 000 s.

P6.55 The ball launcher in a pinball machine has a spring that has a force constant of 1.20 N/cm (Fig. P6.55). The surface on which the ball moves is inclined 10.0° with respect to the horizontal. The spring is initially compressed 5.00 cm. Find the launching speed of a 100-g ball when the plunger is released. Friction and the mass of the plunger are negligible.

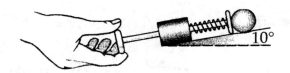

FIG. P6.55

Solution Use the work-kinetic energy theorem. $W_{\text{net}} = \Delta K$: $\qquad K_i + W_s + W_g = K_f$

$$0 + \frac{1}{2}kx^2 - mg\Delta r \sin 10.0° = \frac{1}{2}mv_f^2.$$

Since $v_i = 0$, $v_f = \sqrt{\dfrac{2}{m}\left[\dfrac{1}{2}kx^2 - mg\Delta r \sin 10.0°\right]}$. In this case,

$x = \Delta r = 5.00 \text{ cm} = 5.00 \times 10^{-2} \text{ m}$ and $k = 1.20 \text{ N/cm} = 120 \text{ N/m}$

$$v_f = \sqrt{\frac{2\left[\frac{1}{2}(120 \text{ N/m})(5.00 \times 10^{-2} \text{ m})^2 - (0.100 \text{ kg})(9.80 \text{ m/s}^2)(5.00 \times 10^{-2} \text{ m})(\sin 10.0°)\right]}{0.100 \text{ kg}}}$$

and $v_f = 1.68 \text{ m/s}$. ◊

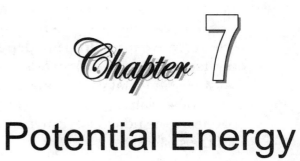

Potential Energy

Section 7.1   Potential Energy of a System

The work done on an object by the force of gravity is equal to the object's initial potential energy minus its final potential energy. The gravitational potential energy associated with an object depends only on the object's weight and its vertical height above the surface of the Earth. If the height above the surface increases, the potential energy will also increase; but the work done by the gravitational force will be negative. In working problems involving gravitational potential energy, it is necessary to choose an arbitrary reference level (or location) at which the potential energy is zero.

Section 7.2   The Isolated System

In an isolated system, the total mechanical energy (sum of the kinetic and potential energies) remains constant. In equation form this is a statement of conservation of mechanical energy, $\Delta K + \Delta U = 0$. This condition holds when the only forces doing the work on the system are conservative forces.

Section 7.3   Conservative and Nonconservative Forces

A force is said to be conservative if the work done by that force on a body moving between any two points is independent of the path taken. In addition, the work done by a conservative force is zero when the body moves through any closed path and returns to its initial position. Nonconservative forces are those for which the work done on a particle moving between two points depends on the path. Furthermore, the work done by a nonconservative force (e.g. friction) around a closed path is not zero.

Conservative forces among members of a system cause no transfer of mechanical energy to internal energy within the system. The work done by a conservative force does not depend upon the path followed by members of the system; all the work depends only on the initial and final configurations of the system. The total work done by all forces acting on a system (conservative and nonconservative) equals the change in the kinetic energy of the system. The work done by the nonconservative forces equals the change in the mechanical energy of the system.

Section 7.4   Conservative Forces and Potential Energy

It is possible to define a potential energy function associated with a conservative force such that the work done by the conservative force equals the negative of the change in the potential energy associated with the force.

Section 7.5   The Nonisolated System in Steady State

In an **isolated** system, no energy transfer occurs across the system boundary; the energy associated with a **nonisolated** system changes due to energy transfers across the boundary. When the nonisolated system is in steady state, the rates of energy entering and leaving the system are equal; the total energy of the system remains constant.

Section 7.7   Energy Diagrams and Stability of Equilibrium

Positions of stable equilibrium correspond to points for which $U(x)$ has a relative minimum value on an energy diagram. Positions of unstable equilibrium correspond to those points for which $U(x)$ has a relative maximum value on an energy diagram. Finally, a position of neutral equilibrium corresponds to a region over which $U(x)$ remains constant.

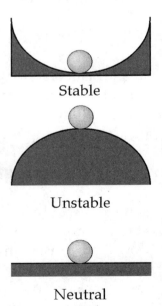

Stable

Unstable

Neutral

FIG. 7.1 Examples of stable, unstable, and neutral equilibrium.

EQUATIONS AND CONCEPTS

The **gravitational potential energy** associated with an object at any point in space is the product of the object's weight and the vertical coordinate relative to an arbitrary reference level.

$$U_g \equiv mgy \tag{7.2}$$

In calculating the **work done by the gravitational force**, remember that the difference in potential energy between two points is independent of the location of the origin. Choose an origin that is convenient to calculate U_i and U_f for a particular situation. *The work done on any system by the gravitational force is equal to the negative of the change in the gravitational potential energy.*

$$W_g = -\left(U_f - U_i\right) = -\Delta U_g$$

The units of energy (kinetic and potential) are the same as the units of work—joules.

$$1\,J = 1\,N \cdot m$$

The law of **conservation of mechanical energy** states that if only conservative forces act on a system, the sum of the kinetic and potential energies remains constant, or is conserved. *According to this important conservation law, if the kinetic energy of the system increases by some amount, the potential energy must decrease by the same amount--and vice versa.* If more than one conservative force acts on an object, then a potential energy term is associated with each force.

$$K_i + U_i = K_f + U_f \tag{7.7}$$

If the **only the gravitational force** is present, the equation for conservation of mechanical energy takes a special form.

$$\tfrac{1}{2}mv_i^2 + mgy_i = \tfrac{1}{2}mv_f^2 + mgy_f$$

Conservation of mechanical energy for a **mass-spring system** is similar:

$$\tfrac{1}{2}mv_i^2 + \tfrac{1}{2}kx_i^2 = \tfrac{1}{2}mv_f^2 + \tfrac{1}{2}kx_f^2$$

Note that both the gravitational force and the spring force satisfy the required properties of a conservative force; the work done is path independent and is zero for any closed path.

Total mechanical energy of a system is the sum of the kinetic and potential energies.	$E_{\text{mech}} \equiv K + U$	(7.8)

An **elastic potential energy function** is associated with a mass-spring system that has been deformed a distance x from the equilibrium position. The force constant k has units of N/m and is characteristic of a particular spring. *The elastic potential energy of a deformed spring is always positive.*

$$U_s \equiv \frac{1}{2}kx^2 \qquad (7.9)$$

When **nonconservative forces** (such as friction) act within a system, the result is a change in the mechanical energy of the system.

$$-f_k d = \Delta K + \Delta U \qquad (7.10)$$

This form of the **continuity equation for energy** is a starting point for solving a wide class of mechanical problems. The first three terms on the left-hand side of the equation refer to the energy of a system at a chosen time and location (referred to as the "initial" point, designated by i). The work and friction terms on the left-hand side of the equation account for the changes in the mechanical energy of a nonisolated system. The terms on the right-hand side of the equation add to give the total energy at a "final" time and location designated by f.

$$K_i + U_{gi} + U_{si} + W_{\text{external}} - f_k d$$
$$= K_f + U_{gf} + U_{sf}$$

A **potential energy function** U can be defined for a conservative force such that the work done by a conservative force equals the negative of the change in the potential energy of the system. *The work done on an object by a conservative force depends only on the initial and final positions of the object and equals zero around a closed path. If more than one conservative force acts, then a potential energy function is associated with each force.*

$$U_f = -\int_{x_i}^{x_f} F_x dx + U_i \qquad (7.15)$$

There is an important **relationship between a conservative force and the potential energy** of the system on which the force acts. *The x component of a conservative force equals the negative derivative of the system's potential energy with respect to x.*

$$F_x = -\frac{dU}{dx} \qquad (7.16)$$

The **gravitational potential energy** associated with a system of two particles or uniform spherical distributions of mass separated by a distance *r* is similar in form to the expression for the electrical potential energy of a system of two charges. *Note that the gravitational potential is negative (the gravitational force is always attractive and the zero level is taken at r = infinity) while the electrical force can be attractive or repulsive.*

$$U_g = -\frac{Gm_1m_2}{r} \qquad (7.20)$$

$$U_e = k_e\frac{q_1q_2}{r} \qquad (7.23)$$

SUGGESTIONS, SKILLS, AND STRATEGIES

Choosing a Zero Level

In working problems involving gravitational potential energy, it is always necessary to choose a location at which the gravitational potential energy is zero. This choice is completely arbitrary because the important quantity is the difference in potential energy, and that difference is independent of the location of zero. It is often convenient, but not essential, to choose the surface of the Earth as the reference position for zero potential energy. In most cases, the statement of the problem suggests a convenient level to use.

Conservation of Energy

Take the following steps in applying the principle of conservation of energy:

- Define your system, which may consist of more than one object. Determine if any energy transfers occur across the boundary. If so, use the nonisolated system model $\Delta E_{system} = \sum T$. If not, use the isolated system model $\Delta E_{system} = 0$.

- Select a reference position for the zero point of each form of potential energy.

- Determine whether or not nonconservative forces (e.g. friction or air resistance) are present.

- If mechanical energy is conserved (that is, if only conservative forces are present), you can write the total initial energy at some point as the sum of the kinetic and potential energies at that point. Then, write an expression for the total final energy, $K_f + U_f$, at the final point of interest. Since mechanical energy is conserved, you can equate the two total energies and solve for the unknown.

- If nonconservative forces such as friction are present, mechanical energy is not conserved. Write expressions for the total initial and total final energies. The difference between the two total energies is equal to the mechanical energy gained or lost due to the presence of nonconservative forces.

REVIEW CHECKLIST

✓ Recognize that the gravitational potential energy function, $U_g = mgy$, can be positive, negative, or zero, depending on the location of the reference level used to measure y. Be aware of the fact that although U depends on the origin of the coordinate system, *the change in potential energy, $U_f - U_i$, is independent of the coordinate system used to define U.*

✓ Understand that a force is said to be conservative if the work done by that force on a body moving between any two points is independent of the path taken. Nonconservative forces are those for which the work done on a particle moving between two points depends on the path. Account for nonconservative forces acting on a system using the work-energy theorem. In this case, the work done by all nonconservative forces equals the change in total mechanical energy of the system.

✓ Understand the distinction between kinetic energy (energy associated with motion), potential energy (energy associated with the position or coordinates of a system), and the total mechanical energy of a system. State the law of conservation of mechanical energy, noting that mechanical energy is conserved when only conservative forces act on a system. This extremely powerful concept is important in all areas of physics.

ANSWERS TO SELECTED QUESTIONS

Q7.3   One person drops a ball from the top of a building, while another person at the bottom observes its motion. Will these two people agree on the value of the gravitational potential energy of the ball-Earth system? On the change in potential energy? On the kinetic energy?

Answer   The two will not necessarily agree on the potential energy, since this depends on the origin—which may be chosen differently for the two observers. However, the two **must** agree on the value of the **change** in potential energy, which is independent of the choice of the reference frames. The two will also agree on the kinetic energy of the ball, assuming both observers are at rest with respect to each other, and hence measure the same v.

Q7.9   You ride a bicycle. In what sense is your bicycle solar-powered?

Answer   The energy to ride the bicycle comes from your body. The source of that energy is the food that you ate at some previous time. The energy in the food, assuming that we focus on vegetables, came from the growth of the plant, for which photosynthesis is a major factor. The light for the photosynthesis comes from the Sun. The argument for meats has a couple of extra steps, but also goes through the process of photosynthesis in the plants eaten by animals. Thus, the source of the energy to ride the bicycle is the Sun, and your bicycle is solar-powered!

Q7.11   A bowling ball is suspended from the ceiling of a lecture hall by a strong cord. The ball is drawn away from its equilibrium position and released from rest at the tip of the demonstrator's noise as shown in Figure Q7.11. If the demonstrator remains stationary, explain why the ball will not strike her on its return swing. Would this demonstrator be safe if the ball were given a push from its starting position at her nose?

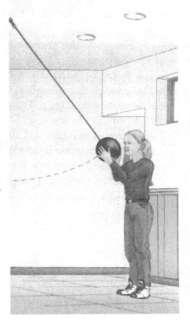

Answer   The total energy of the system (the bowling ball and the Earth) must be conserved. Since the system initially has a potential energy mgh, and the ball has no kinetic energy, it cannot have any kinetic energy when returning to its initial position. Of course, air resistance will cause the ball to return to a point slightly below its initial position. On the other hand, if the ball is given a push, the demonstrator's nose will be in big trouble.

FIG. Q7.11

SOLUTIONS TO SELECTED PROBLEMS

P7.5 A bead slides without friction around a loop-the-loop (Fig. P7.5). The bead is released from a height $h = 3.50R$.

(a) What is its speed at point A?

(b) How large is the normal force on it if its mass is 5.00 g?

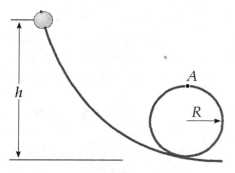

FIG. P7.5

Solution Conceptualize: Since the bead is released above the top of the loop, it will have enough potential energy to reach point A and still have excess kinetic energy. The energy of the bead at point A will be proportional to h and g. If it is moving relatively slowly, the track will exert an upward force on the bead, but if it is whipping around fast, the normal force will push it toward the center of the loop.

Categorize: The speed at the top can be found from the conservation of energy, and the normal force can be found from Newton's second law.

Analyze:

(a) We define the bottom of the loop as the zero level for the gravitational potential energy.

Since $v_i = 0$, $$E_i = K_i + U_i = 0 + mgh = mg(3.50R).$$

The total energy of the bead at point A can be written as

$$E_A = K_A + U_A = \frac{1}{2}mv_A^2 + mg(2R).$$

Since mechanical energy is conserved, $E_i = E_A$

and we get $$\frac{1}{2}mv_A^2 + mg(2R) = mg(3.50R),$$

$$v_A^2 = 3.00gR,$$

or $v_A = \sqrt{3.00gR}$. ◊

continued on next page

(b) To find the normal force at the top, we may construct a free-body diagram as shown, where we assume that $\vec{\mathbf{n}}$ is downward, like $m\vec{\mathbf{g}}$. Newton's second law gives $\sum F = ma_c$, where a_c is the centripetal acceleration.

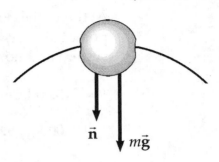

$$n + mg = \frac{mv_A{}^2}{R} = \frac{m(3.00gR)}{R} = 3.00mg$$

$$n = 3.00mg - mg = 2.00mg$$

$$n = 2.00\left(5.00 \times 10^3 \text{ kg}\right)\left(9.80 \text{ m/s}^2\right) = 0.098\ 0 \text{ N downward}$$

FIG. P7.5(b)

Finalize: Our answer represents the speed at point *A* as proportional to the square root of the product of *g* and *R*, but we must not think that simply increasing the diameter of the loop will increase the speed of the bead at the top. In general, the speed will increase with increasing release height, which for this problem was defined in terms of the radius. The normal force may seem small, but it is twice the weight of the bead.

P7.9 Two objects are connected by a light string passing over a light, frictionless pulley as shown in Figure P7.9. The 5.00-kg object is released from rest. Using the principle of conservation of energy,

(a) determine the speed of the 3.00-kg object just as the 5.00-kg object hits the ground, and

(b) find the maximum height to which the 3.00-kg object rises.

Solution As the system choose the two blocks *A* and *B*, string, pulley, and Earth.

(a) Choose the initial point before release and the final point just before the larger object hits the floor. The total energy of the system remains constant and the energy version of the isolated system model gives

$$(K_A + K_B + U_g)_i = (K_A + K_B + U_g)_f$$

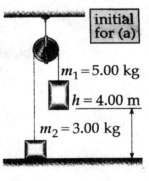

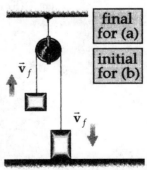

FIG. P7.9(a)

continued on next page

At the initial point K_{Ai} and K_{Bi} are zero and we define the gravitational potential energy of the system as zero. No external forces do work on the system and no friction acts within the system. Thus,

$$0 = \frac{1}{2}(5.00 \text{ kg} + 3.00 \text{ kg})v_f^2 + (3.00 \text{ kg})(9.80 \text{ m/s}^2)(4.00 \text{ m})$$

$$+(5.00 \text{ kg})(9.80 \text{ m/s}^2)(-4.00 \text{ m})$$

$$(2.00 \text{ kg})(9.80 \text{ m/s}^2)(4.00 \text{ m}) = \frac{1}{2}(8.00 \text{ kg})v_f^2$$

$$v_f = 4.43 \text{ m/s} \qquad \lozenge$$

(b)   Now the string goes slack. The 3.00-kg object becomes a projectile. We focus now on the system of the 3.00-kg object and the Earth. Take the initial point at the previous final point, and the new final point at its maximum height:

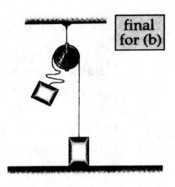

FIG. P7.9(b)

$$(K + U_g)_i = (K + U_g)_f$$

$$\frac{1}{2}(3.00 \text{ kg})(4.43 \text{ m/s})^2 + (3.00 \text{ kg})(9.80 \text{ m/s}^2)(4.00 \text{ m})$$

$$= 0 + (3.00 \text{ kg})(9.80 \text{ m/s}^2)y_f$$

$$y_f = 5.00 \text{ m} \qquad \lozenge$$

or 1.00 m higher than the height of the 5.00-kg mass when it was released.

P7.15   A force acting on a particle moving in the xy plane is given by $\vec{F} = \left(2y\hat{i} + x^2\hat{j}\right)$ N, where x and y are in meters. The particle moves from the origin to a final position having coordinates $x = 5.00$ m and $y = 5.00$ m, as in Figure P7.15. Calculate the work done by $\vec{F}$ on the particle as it moves along

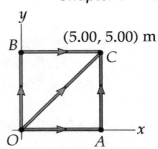

FIG. P7.15

(a)   *OAC*

(b)   *OBC*

(c)   *OC.*

(d)   Is $\vec{F}$ conservative or nonconservative? Explain.

Solution   In the following integrals, remember $\hat{i} \cdot \hat{i} = \hat{j} \cdot \hat{j} = 1$ and $\hat{i} \cdot \hat{j} = 0$.

(a) $W_{OA} = \int_0^{5.00} \left(2y\hat{i} + x^2\hat{j}\right) \cdot \left(\hat{i}\,dx\right) = \int_0^{5.00} 2y\,dx = 2y \int_0^{5.00} dx = 2yx \Big]_{x=0,y=0}^{x=5.00,y=0} = 0$

$W_{AC} = \int_0^{5.00} \left(2y\hat{i} + x^2\hat{j}\right) \cdot \left(\hat{j}\,dy\right) = \int_0^{5.00} x^2\,dy = x^2 \int_0^{5.00} dy = x^2 y \Big]_{x=5.00,y=0}^{x=5.00,y=5.00} = 125 \text{ J}$

$W_{OAC} = 0 + 125 \text{ J} = 125 \text{ J}$ ◊

(b) $W_{OB} = \int_0^{5.00} \left(2y\hat{i} + x^2\hat{j}\right) \cdot \left(\hat{j}\,dy\right) = \int_0^{5.00} x^2\,dy = x^2 \int_0^{5.00} dy = x^2 y \Big]_{x=0...}^{x=0...} = 0$

$W_{BC} = \int_0^{5.00} \left(2y\hat{i} + x^2\hat{j}\right) \cdot \left(\hat{i}\,dx\right) = \int_0^{5.00} 2y\,dx = 2y \int_0^{5.00} dx = 2(5.00)x \Big]_0^{5.00} = 50.0 \text{ J}$

$W_{OBC} = 0 + 50.0 \text{ J} = 50.0 \text{ J}$ ◊

(c) $W_{OC} = \int \left(2y\hat{i} + x^2\hat{j}\right) \cdot \left(\hat{i}\,dx + \hat{j}\,dy\right) = \int_{x=0,y=0}^{x=5.00,y=5.00} \left(2y\,dx + x^2\,dy\right)$

Since $x = y$ along OC, $dx = dy$ and $W_{OC} = \int_0^{5.00} \left(2x + x^2\right)dx = 66.7 \text{ J}$ ◊

(d) $\vec{F}$ is non-conservative since the work done is path dependent.   ◊

P7.17   A block of mass 0.250 kg is placed on top of a light vertical spring of force constant 5 000 N/m and is pushed downward so that the spring is compressed 0.100 m. After the block is released from rest, it travels upward and then leaves the spring. To what maximum height above the point of release does it rise?

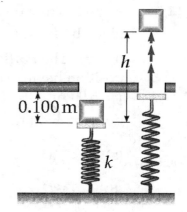

FIG. P7.17

Solution   In both the initial and final states, the block is not moving. Therefore, the initial and final energies of the block-spring-Earth system are:

$$E_i = K_i + U_i = 0 + \left(U_g + U_s\right)_i = 0 + \left(0 + \frac{1}{2}kx^2\right)$$
$$E_f = K_f + U_f = 0 + \left(U_g + U_s\right)_f = 0 + \left(mgh + 0\right)$$

Since $E_i = E_f$, $\qquad\qquad\qquad\qquad mgh = \frac{1}{2}kx^2$

and $h = \dfrac{kx^2}{2mg} = \dfrac{(5000 \ \text{N/m})(0.100 \ \text{m})^2}{2(0.250 \ \text{kg})(9.80 \ \text{m/s}^2)} = 10.2 \ \text{m}$ ◊

P7.23   The coefficient of friction between the 3.00-kg block and surface in Figure P7.23 is 0.400. The system starts from rest. What is the speed of the 5.00-kg ball when it has fallen 1.50 m?

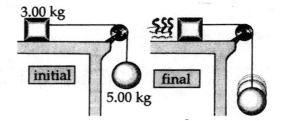

FIG. P7.23

Solution   **Conceptualize:** Assuming that the block does not reach the pulley within the 1.50 m distance, a reasonable speed for the ball might be somewhere between 1 and 10 m/s based on common experience.

Categorize: We could solve this problem by using $\sum F = ma$ to give a pair of simultaneous equations in the unknown acceleration and tension; then we would have to solve a motion problem to find the final speed. We may find it easier to solve using the work-energy theorem.

continued on next page

Analyze: For the Earth plus objects A (block) and B (ball), the work-kinetic energy theorem is

$$\left(K_A + K_B + U_A + U_B\right)_i + W_{app} - f_k d = \left(K_A + K_B + U_A + U_B\right)_f.$$

Choose the initial point before release and the final point after each block has moved 1.50 m. Choose $U_g = 0$ with the 3.00-kg block on the tabletop and the 5.00-kg block in its final position. So

$$K_{Ai} = K_{Bi} = U_{Ai} = U_{Af} = U_{Bf} = 0.$$

Also, since the only external forces are gravity and friction,

$$W_{app} = 0.$$

We now have

$$0 + 0 + 0 + m_B g y_{Bi} + 0 - f_k d = \frac{1}{2} m_A v_f^2 + \frac{1}{2} m_B v_f^2 + 0 + 0$$

where the frictional force is

$$f_k = \mu_k n = \mu_k m_A g$$

and causes a negative change in mechanical energy since the force opposes the motion. Since all of the variables are known except for v_f, we can substitute and solve for the final speed.

$$(5.00 \text{ kg})(9.80 \text{ m/s}^2)(1.50 \text{ m}) - (0.400)(3.00 \text{ kg})(9.80 \text{ m/s}^2)(1.50 \text{ m}) = \ldots$$

$$\ldots = \frac{1}{2}(3.00 \text{ kg})v_f^2 + \frac{1}{2}(5.00 \text{ kg})v_f^2$$

$$73.5 \text{ J} - 17.6 \text{ J} = \frac{1}{2}(8.00 \text{ kg})v_f^2$$

or $v_f = \sqrt{\dfrac{2(55.9 \text{ J})}{8.00 \text{ kg}}} = 3.74 \text{ m/s}$ ◊

Finalize: The final speed seems reasonable based on our expectation. This speed must also be less than if the rope were cut and the ball simply fell, in which case its final speed would be

$$v_f' = \sqrt{2gy} = \sqrt{2(9.80 \text{ m/s}^2)(1.50 \text{ m})} = 5.42 \text{ m/s}$$

P7.25 A 5.00-kg block is set into motion up an inclined plane with an initial speed of 8.00 m/s (Fig. P7.25). The block comes to rest after traveling 3.00 m along the plane, which is inclined at an angle of 30.0° to the horizontal.

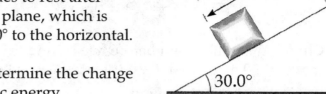

(a) For this motion, determine the change in the block's kinetic energy

(b) For this motion, determine the change in potential energy of the block-Earth system

FIG. P7.25

(c) For this motion, determine the friction force exerted on the block (assumed to be constant).

(d) What is the coefficient of kinetic friction?

Solution (a) $\Delta K = K_f - K_i = \frac{1}{2}mv_f^2 - \frac{1}{2}mv_i^2 = 0 - \frac{1}{2}(5.00 \text{ kg})(8.00 \text{ m/s})^2$

$= -160 \text{ J}$ ◊

(b) $\Delta U = U_{gf} - U_{gi} = mgy_f - 0 = (5.00 \text{ kg})(9.80 \text{ m/s}^2)(3.00 \text{ m}) \sin 30.0° = 73.5 \text{ J}$ ◊

(c) $(K + U_g)_i + W_{external} - f_k d = (K + U_g)_f$

$\frac{1}{2}(5.00 \text{ kg})(8.00 \text{ m/s})^2 + 0 - f(3.00 \text{ m}) = 0 + (5.00 \text{ kg})(9.80 \text{ m/s}^2)(1.50 \text{ m})$

$160 \text{ J} - f(3.00 \text{ m}) = 73.5 \text{ J}$ and $f = \frac{86.5 \text{ J}}{3.00 \text{ m}} = 28.8 \text{ N}$ ◊

(d) The forces perpendicular to the incline must add to zero.

$\sum F_y = 0$: $+n - mg\cos 30.0° = 0$.

Substituting, $n = (5.00 \text{ kg})(9.80 \text{ m/s}^2)\cos 30.0° = 42.4 \text{ N}$.

Now, $f_k = \mu_k n$ gives $\mu_k = \frac{f_k}{n} = \frac{28.8 \text{ N}}{42.4 \text{ N}} = 0.679$. ◊

P7.31 A single conservative force acts on a 5.00-kg particle. The equation $F_x = (2x + 4)$ N describes the force, where x is in meters. As the particle moves along the x-axis from $x = 1.00$ m to $x = 5.00$ m, calculate

(a)   the work done by this force.

(b)   the change in the potential energy of the system.

(c)   the kinetic energy of the particle at $x = 5.00$ m if its speed is 3.00 m/s at $x = 1.00$ m.

Solution (a) $W_F = \int_{x_i}^{x_f} F_x \, dx$ where

$$F_x = (2x + 4) \text{ N}, \ x_i = 1.00 \text{ m, and } x_f = 5.00 \text{ m.}$$

Therefore, $W_F = \int_{1.00 \text{ m}}^{5.00 \text{ m}} (2x + 4) dx \text{ N} \cdot \text{m} = x^2 + 4x \Big]_{1.00 \text{ m}}^{5.00 \text{ m}} \text{ N} \cdot \text{m} = 40.0 \text{ J.}$ ◊

(b)   The change in potential energy equals the negative of the work done by the conservative force.

$$\Delta U = -W_F = -40.0 \text{ J}$$ ◊

(c)   When only conservative forces act, conservation of energy gives

$$K_i + U_i = K_f + U_f.$$

Rearranging, $K_f = K_i - \left(U_f - U_i \right) = \frac{1}{2} m v_i^2 - \Delta U$

so $K_f = \frac{1}{2}(5.00 \text{ kg})(3.00 \text{ m/s})^2 - (-40.0 \text{ J}) = 62.5 \text{ J.}$ ◊

P7.33 The potential energy of a system of two particles separated by a distance r is given by $U(r) = \dfrac{A}{r}$, where A is a constant. Find the radial force $\vec{\mathbf{F}}$ that each particle exerts on the other.

Solution The force is the negative derivative of the potential energy with respect to distance:

$$F_r = -\frac{dU}{dr} = -\frac{d}{dr}\left(Ar^{-1} \right) = -A(-1)r^{-2} = \frac{A}{r^2}.$$ ◊

This describes an inverse-square-law force of repulsion, as between two negative point electric charges if A is positive.

P7.49 The particle described in Problem 7.48 (Fig. P7.48) is released from rest at A, and the surface of the bowl is rough. The speed of the particle at B is 1.50 m/s.

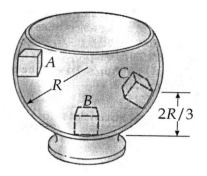

(a) What is its kinetic energy at B?

(b) How much mechanical energy is transformed into internal energy as the particle moves from A to B?

FIG. P7.48

(c) Is it possible to determine the coefficient of friction from these results in any simple manner? Explain.

Solution Let us take $U = 0$ for the particle-bowl-Earth system when the particle is at B. Since $v_i = 0$ at A, $K_A = 0$ and $U_A = mgR$.

(a) Since $v_B = 1.50$ m/s and $m = 200$ g,

$$K_B = \frac{1}{2}mv_B{}^2 = \frac{1}{2}(0.200 \text{ kg})(1.50 \text{ m/s})^2 = 0.225 \text{ J}. \qquad \Diamond$$

(b) At A, $E_i = K_A + U_A = 0 + mgR = (0.200 \text{ kg})(9.80 \text{ m/s}^2)(0.300 \text{ m}) = 0.588 \text{ J}.$

At B, $E_f = K_B + U_B = 0.225 \text{ J} + 0$. The decrease in mechanical energy is equal to the increase in internal energy.

$$E_{\text{mech},\, i} - \Delta E_{\text{int}} = E_{\text{mech},\, f}.$$

The energy transformed is

$$\Delta E_{\text{int}} = -\Delta E_{\text{mech}} = E_i - E_f = 0.588 \text{ J} - 0.225 \text{ J} = 0.363 \text{ J} \qquad \Diamond$$

(c) Even though the energy transformed is known, both the normal force and the friction force change with position as the block slides on the inside of the bowl. Therefore, there is no easy way to find the coefficient of friction. $\qquad \Diamond$

P7.51   A 10.0-kg block is released from point *A* in Figure P7.51. The track is frictionless except for the portion between points *B* and *C*, which has a length of 6.00 m. The block travels down the track, hits a spring of force constant $k = 2\,250$ N/m, and compresses the spring 0.300 m from its equilibrium position before coming to rest momentarily. Determine the coefficient of kinetic friction between the block and the rough surface between *B* and *C*.

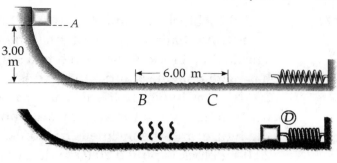

FIG. P7.51

Solution   **Conceptualize:** We should expect the coefficient of friction to be somewhere between 0 and 1 since this is the range of typical μ_k values. It is possible that μ_k could be greater than 1, but it can never be less than 0.

Categorize: The easiest way to solve this problem is by considering the energy changes experienced by the block between the point of release (initial) and the point of full compression of the spring (final). Recall that the change in potential energy (gravitational and elastic) plus the change in kinetic energy must equal the work done on the block by non-conservative forces. Choose the gravitational energy to be zero along the flat portion of the track.

Analyze: Putting the energy equation into symbols: $K_A + U_{gA} - f_k d_{BC} = K_D + U_{sD}$.

Expanding into specific variables: $0 + mgy_A - f_k d_{BC} = 0 + \dfrac{1}{2}kx_s^2$. The friction force is

$f_k = \mu_k mg$, $mgy_A - \dfrac{1}{2}kx^2 = \mu_k mgd$. Solving for the unknown variable μ_k:

$\mu_k = \dfrac{y_A}{d} - \dfrac{kx^2}{2mgd}$. Substituting:

$$\mu_k = \frac{3.00 \text{ m}}{6.00 \text{ m}} - \frac{(2\,250 \text{ N/m})(0.300 \text{ m})^2}{2(10.0 \text{ kg})(9.80 \text{ m/s}^2)(6.00 \text{ m})} = 0.328 \qquad \lozenge$$

Finalize: Our calculated value seems reasonable based on the text's tabulation coefficients of friction. The most important aspect to solving these energy problems is considering how the energy is transferred from the initial to final energy states and remembering to subtract the energy resulting from any non-conservative forces (like friction).

P7.53   A 20.0-kg block is connected to a 30.0-kg block by a string that passes over a frictionless pulley. The 30.0-kg block is connected to a spring that has negligible mass and a force constant of 250 N/m, as in Figure P7.53. The spring is unstretched when the system is as shown in the figure, and the incline is frictionless. The 20.0-kg block is pulled 20.0 cm down the incline (so that the 30.0-kg block is 40.0 cm above the floor) and released from rest. Find the speed of each block when the 30.0-kg block is again 20.0 cm above the floor (i.e., when the spring is unstretched).

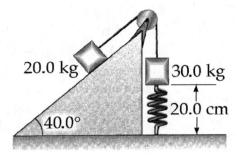

FIG. P7.53

Solution   Let x be the distance the spring is stretched from equilibrium ($x = 0.200$ m), which corresponds to the upward displacement of the 30.0-kg mass. Also let $U_g = 0$ be measured with respect to the lowest position of the 20.0-kg mass when the system is released from rest. Finally, define v as the speed of both blocks at the moment the spring passes through its unstretched position. Since all forces are conservative, conservation of energy yields $\Delta K + \Delta U_s + \Delta U_g = 0$. Solving for each variable,

$$\Delta K = \frac{1}{2}(m_1 + m_2)v^2 - 0 = \frac{1}{2}(50.0 \text{ kg})v^2 = (25.0 \text{ kg})v^2$$

$$\Delta U_s = 0 - \frac{1}{2}kx^2 = -\frac{1}{2}(250 \text{ N/m})(0.200 \text{ m})^2 = -5.00 \text{ N} \cdot \text{m}$$

$$\Delta U_g = (m_2 \sin\theta - m_1)gx = \left[(20.0 \text{ kg})\sin 40.0° - 30.0\text{kg}\right](9.80 \text{ m/s}^2)(0.200 \text{ m})$$
$$= -33.6 \text{ N} \cdot \text{m}$$

Substituting into our equation representing the energy version of the isolated system model,

$$(25.0 \text{ kg})v^2 - 5.00 \text{ N} \cdot \text{m} - 33.6 \text{ N} \cdot \text{m} = 0.$$

Solving for v gives $v = 1.24$ m/s.   ◊

P7.55 A block of mass 0.500 kg is pushed against a horizontal spring of negligible mass until the spring is compressed a distance x (Fig. P7.55). The force constant of the spring is 450 N/m. When released, the block travels along a frictionless, horizontal surface to point B, the bottom of a vertical circular track of radius $R = 1.00$ m, and continues to move up the track. The speed of the block at the bottom of the track is $v_B = 12.0$ m/s, and the block experiences an average frictional force of 7.00 N while sliding up the track.

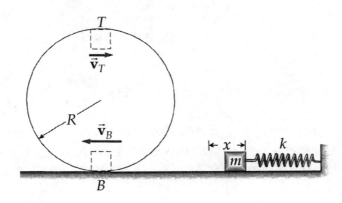

FIG. P7.55

(a) What is x?

(b) What speed do you predict for the block at the top of the track?

(c) Does the block reach the top of the track, or does it fall off before reaching the top?

Solution The energy of the block-spring system is conserved in the firing of the block. Therefore,

(a) $\dfrac{1}{2}kx^2 = \dfrac{1}{2}mv^2$ or $\dfrac{1}{2}(450\ \text{N/m})(x)^2 = \dfrac{1}{2}(0.500\ \text{kg})(12.0\ \text{m/s})^2.$

Thus, $x = 0.400$ m. ◊

(b) To find speed of block at the top, we consider the block-Earth system.

$$(K+U_g)_B - f_k d = (K+U_g)_T : \left(mgh_B + \frac{1}{2}mv_B{}^2 \right) - f(\pi R) = \left(mgh_T + \frac{1}{2}mv_T{}^2 \right)$$

continued on next page

Substituting, $\qquad mgh_T = (0.500 \text{ kg})(9.80 \text{ m/s}^2)(2.00 \text{ m}) = 9.80 \text{ J}.$

We have

$$0 + \frac{1}{2}(0.500 \text{ kg})(12.0 \text{ m/s})^2 - (7.00 \text{ N})(\pi)(1.00 \text{ m}) = 9.80 \text{ J} + \frac{1}{2}(0.500 \text{ kg})v_T^2$$

and $\qquad 0.250 v_T{}^2 = 4.21.$

Thus, $\qquad v_T = 4.10 \text{ m/s}.$ $\qquad\qquad$ ◊

(c) The block falls if $a_c < g$ $\qquad a_c = \dfrac{v_T{}^2}{R} = \dfrac{(4.10 \text{ m/s})^2}{1.00 \text{ m}} = 16.8 \text{ m/s}^2.$

Therefore, $a_c > g$. Some downward normal force is required along with the block's weight to provide the centripetal acceleration, and the block stays on the track. $\qquad\qquad$ ◊

Chapter 8

Momentum and Collisions

Section 8.1   Linear Momentum and Its Conservation

Section 8.2   Impulse and Momentum

If two particles of masses m_1 and m_2 form an isolated system, then the total momentum of the system remains constant. The time rate of change of the momentum of a particle is equal to the resultant force on the particle. The impulse of a force equals the change in momentum of the particle on which the force acts. Under the impulse approximation, it is assumed that one of the forces acting on a particle is of short time duration but of much greater magnitude than any of the other forces.

Section 8.3   Collisions

For **any type of collision**, the total momentum before the collision equals the total momentum just after the collision.

In an **elastic collision**, both momentum and kinetic energy are conserved.

In an **inelastic collision**, the total momentum is conserved; however, the total kinetic energy is not conserved.

In a **perfectly inelastic collision**, the two colliding objects stick together following the collision.

Section 8.4   Two-Dimensional Collisions

The law of conservation of momentum is not restricted to one-dimensional collisions. If two masses undergo a two-dimensional (glancing) collision and there are no external forces acting, the total momentum in each of the x, y, and z directions is conserved.

Section 8.5   The Center of Mass

The position of the center of mass of a system can be described as the average position of the system's mass. If g is constant over a mass distribution, then the center of gravity will coincide with the center of mass.

Section 8.6   Motion of a System of Particles

The center of mass of a system of particles moves like an imaginary particle of mass M (equal to the total mass of the system) under the influence of the resultant external force on the system.

Section 8.7   Rocket Propulsion

The operation of a rocket depends on the law of conservation of linear momentum applied to a system of particles (the rocket plus its ejected fuel). The thrust of the rocket is the force exerted on it by the exhaust gases.

EQUATIONS AND CONCEPTS

The **linear momentum $\vec{p}$** of a particle is defined as the product of its mass m with its velocity, $\vec{v}$.

$$\vec{p} = m\vec{v} \tag{8.2}$$

Equation 8.2 is equivalent to three component scalar equations, one along each of the coordinate axes.

$$p_x = mv_x \quad p_y = mv_y \quad p_z = mv_z \tag{8.3}$$

The **net force** acting on a particle is equal to the time rate of change of the linear momentum of the particle. *This is an alternative form of Newton's second law.*

$$\sum \vec{F} = \frac{d\vec{p}}{dt} \tag{8.4}$$

The **law of conservation of momentum** states: when the net external force acting on a system of particles is zero, the total linear momentum of the system is conserved. There may be internal forces acting between particles within the system; conservation of momentum only requires that the net external force be zero. *This fundamental law is especially useful in treating problems involving collisions between two bodies.*

$$\vec{p}_{tot} = \text{constant} \tag{8.5}$$

$$\text{when } \sum F_{ext} = 0$$

The **total momentum of an isolated pair of particles** remains constant. This is true because the net external force equals the time rate of change of momentum; and in an isolated system only internal forces are present. *Momentum is a vector quantity and the momentum along each of the coordinate directions is independently conserved.*

$$\vec{p}_{1i} + \vec{p}_{2i} = \vec{p}_{1f} + \vec{p}_{2f} \tag{8.6}$$

The **impulse-momentum theorem** states that the impulse of a force $\vec{F}$ acting on a particle equals the change in the momentum of the particle.

$$\vec{I} = \int_{t_i}^{t_f} \sum \vec{F}_{ext}\, dt = \Delta \vec{P}_{total} \tag{8.11}$$

The time-averaged force, $\vec{F}_{ave}$, is defined as that constant force which would give the same impulse to a particle as an actual time-varying force over the same time interval Δt (see the shaded area in the figure). *The impulse approximation assumes that one of the forces acting on a particle acts for a short time and is much larger than any other force present.* This approximation is usually made in collision problems, where the force of interest is the contact force between the particles during the collision.

$$\sum \vec{F}_{ave} \equiv \frac{1}{\Delta t} \int_{t_i}^{t_f} \sum \vec{F}\, dt \tag{8.12}$$

$$\vec{I} = \sum \vec{F}_{ave}\, \Delta t \tag{8.13}$$

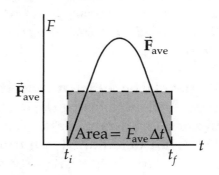

Consider different **types of collisions** that can occur between two bodies:

An elastic collision is one in which both linear momentum and kinetic energy are conserved.

$$\vec{p}_1 + \vec{p}_2 = \text{constant}$$
$$K_1 + K_2 = \text{constant} \quad \text{(Elastic)}$$

An inelastic collision is one in which only linear momentum is conserved.

$$\vec{p}_1 + \vec{p}_2 = \text{constant}$$
$$K_1 + K_2 \neq \text{constant} \quad \text{(Inelastic)}$$

A perfectly inelastic collision is an inelastic collision in which the two bodies stick together and move with a common velocity following the collision.

Note that momentum is conserved in any type of collision. Furthermore, note that when we say that the momentum is conserved, we are speaking about the momentum of the entire system. *The momentum of each particle may change as the result of the collision, but the momentum of the system remains unchanged.*

The **common velocity following a perfectly inelastic collision** (here assumed one-dimensional) between two bodies can be calculated in terms of the two mass values and the two initial velocities.

$$v_f = \frac{m_1 v_{1i} + m_2 v_{2i}}{m_1 + m_2} \tag{8.16}$$

The **relative velocity** before a perfectly elastic collision between two bodies equals the negative of the relative velocity of the two particles following the collision. *When two particles undergo a perfectly elastic collision, both momentum and kinetic energy are conserved.*

$$v_{1i} - v_{2i} = -\left(v_{1f} - v_{2f}\right) \tag{8.21}$$

Following an **elastic collision between two objects**, the final velocities can be found in terms of the initial velocities and masses. The signs of the velocities are determined by the directions of motion. An important special case occurs when the second particle (m_2, the "target") is initially at rest. *Remember the appropriate algebraic signs (designating direction) must be included for v_{1i} and v_{2i}.*

$$v_{1f} = \left(\frac{m_1 - m_2}{m_1 + m_2}\right)v_{1i} + \left(\frac{2m_2}{m_1 + m_2}\right)v_{2i} \tag{8.22}$$

$$v_{2f} = \left(\frac{2m_1}{m_1 + m_2}\right)v_{1i} + \left(\frac{m_2 - m_1}{m_1 + m_2}\right)v_{2i} \tag{8.23}$$

The *x*-**coordinate of the center of mass of** *n* **particles** whose individual coordinates are x_1, x_2, x_3, ... and whose masses are m_1, m_2, m_3, ... is given by Equation 8.30. The *y* and *z* coordinates of the center of mass are defined by similar expressions. *The center of mass of a homogeneous, symmetric body must lie on an axis of symmetry.*

$$x_{CM} \equiv \frac{m_1 x_1 + m_2 x_2 + m_3 x_3 + \ldots + m_n x_n}{m_1 + m_2 + m_3 + \ldots + m_n}$$
$$= \frac{\sum\limits_i m_i x_i}{M} \tag{8.30}$$

The **center of mass for a collection of particles** can be located by its position vector. In Equation 8.32, $\vec{r}_i$ is the position vector of the ith particle.

$$\vec{r}_{CM} = \frac{\sum_i m_i \vec{r}_i}{M} \qquad (8.32)$$

$$\vec{r}_i \equiv x_i \hat{i} + y_i \hat{j} + z_i \hat{k}$$

The **center of mass of an extended object** can be calculated by integrating over the total length, area, or volume which includes the total mass M.

$$\vec{r}_{CM} = \frac{1}{M} \int \vec{r} \, dm \qquad (8.35)$$

The **velocity of the center of mass of a system of particles** is given by Equation 8.36, where $\vec{v}_i$ is the velocity of the ith particle and M is the total mass of the system.

$$\vec{v}_{CM} = \frac{d\vec{r}_{CM}}{dt} = \frac{1}{M} \sum_i m_i \vec{v}_i \qquad (8.36)$$

The **acceleration of the center of mass of a system of particles** depends on the value of the acceleration for each of the individual particles.

$$\vec{a}_{CM} = \frac{d\vec{v}_{CM}}{dt} = \frac{1}{M} \sum_i m_i \vec{a}_i \qquad (8.38)$$

The **time rate of change of the total momentum** of a system of particles depends on the net external force acting on the system. If the net external force on the system is zero, then the total momentum of the system remains constant. *This form of Newton's second law must be used when the mass of the system changes.*

$$\sum \vec{F}_{ext} = M\vec{a}_{CM} = \frac{d\vec{p}_{tot}}{dt} \qquad (8.40)$$

The **total momentum of a system of particles** is equal to the total mass M multiplied by the velocity of the center of mass.

$$\vec{p}_{tot} = M\vec{v}_{CM} = \text{constant} \qquad (8.41)$$

$$\left(\text{when } \sum \vec{F}_{ext} = 0\right)$$

The basic **expression for rocket propulsion** states that the change in speed of the rocket, as the mass decreases from M_i to M_f, is proportional to the exhaust speed of the ejected gases. This is the law of conservation of momentum as applied to the rocket and its ejected fuel. If a rocket moves in the absence of gravity and ejects fuel with an exhaust velocity v_e, its change in velocity is proportional to the exhaust velocity, where M_i and M_f refer to its initial and final mass values for the rocket.

$$v_f - v_i = v_e \ln\left(\frac{M_i}{M_f}\right) \qquad (8.43)$$

$$v_e = \text{exhaust velocity}$$

The **thrust** on a rocket increases as the exhaust speed increases and as the burn rate increases.

$$\text{Instantaneous thrust} = \left| v_e \frac{dM}{dt} \right| \qquad (8.44)$$

SUGGESTIONS, SKILLS, AND STRATEGIES

The following procedure is recommended when dealing with problems involving collisions between two objects:

- Set up a coordinate system and define your velocities with respect to that system. That is, objects moving in the direction selected as the positive direction of the x axis are considered as having a positive velocity and negative if moving in the negative-x direction. It is convenient to have the x-axis coincide with one of the initial velocities.

- In your sketch of the coordinate system, draw all velocity vectors with labels and include all the given information.

- Write expressions for the momentum of each object before and after the collision. (In two-dimensional collision problems, write expressions for the x and y components of momentum before and after the collision.) Remember to include the appropriate signs for the velocity vectors.

- Now write expressions for the total momentum before and total momentum after the collision and equate the two. (For two-dimensional collisions, this expression should be written for the momentum in both the x and y directions.) *It is important to emphasize that it is the momentum of the system (the two colliding objects) that is conserved, not the momentum of the individual objects.*

- If the collision is inelastic, you should then proceed to solve the momentum equations for the unknown quantities.

- If the collision is elastic, kinetic energy is also conserved, so you can equate the total kinetic energy before the collision to the total kinetic energy after the collision. This gives an additional relationship between the various velocities. The conservation of kinetic energy for one-dimensional elastic collisions leads to the expression $v_{1i} - v_{2i} = -(v_{1f} - v_{2f})$, which is often easier to use in solving elastic collision problems than is an expression for conservation of kinetic energy.

REVIEW CHECKLIST

✓   The impulse of a force acting on a particle during some time interval equals the change in momentum of the particle, and the impulse equals the area under the force-time graph.

✓   The momentum of any isolated system (one for which the net external force is zero) is conserved, regardless of the nature of the forces between the masses which compose the system.

✓   There are two types of collisions that can occur between two particles, namely elastic and inelastic collisions. Recognize that a perfectly inelastic collision is an inelastic collision in which the colliding particles stick together after the collision, and hence move as a composite particle.

✓   The conservation of linear momentum applies not only to head-on collisions (one-dimensional), but also to glancing collisions (two- or three-dimensional). For example, in a two-dimensional collision, the total momentum in the x direction is conserved and the total momentum in the y direction is conserved.

✓   The equations for momentum and kinetic energy can be used to calculate the final velocities in a two-body head-on elastic collision; and to calculate the final velocity and the change of kinetic energy in a two-body system for a completely inelastic collision.

ANSWERS TO SELECTED QUESTIONS

Q8.3   If two particles have equal kinetic energies, are their momenta necessarily equal? Explain.

Answer   No, their momenta need not be equal. Equal kinetic energies means that $\dfrac{mv^2}{2}$ is the same for both. If they are moving and their masses are different, then their speeds must also be different. Let the speed of the lighter particle be larger by the factor α. Then v^2 is larger by the factor α^2, and the mass of the lighter particle must be smaller by the factor $\dfrac{1}{\alpha^2}$. The momentum takes the increased speed into account only once, not twice like the kinetic energy. The less massive particles will have a momentum smaller by the factor $\left(\dfrac{1}{\alpha^2}\right)\alpha = \dfrac{1}{\alpha}$.

Q8.7   Explain how linear momentum is conserved when a ball bounces from a floor.

Answer   The ball's downward momentum increases as it accelerates downward. A larger upward momentum change occurs when it touches the floor and rebounds. The outside forces of gravity and the normal force inject impulses to change its momentum.

　　　　If we think of ball-and-Earth-together as our system, these forces are internal and do not change the total momentum. It is conserved, if we neglect the curvature of the Earth's orbit. As the ball falls down, the Earth lurches up to meet it, on the order of 10^{25} times more slowly. Then, ball and Earth bounce off each other and separate. In other words, while you dribble a ball, you also dribble the Earth.

Q8.11   A sharpshooter fires a rifle while standing with the butt of the gun against her shoulder. If the forward momentum of a bullet is the same as the backward momentum of the gun, why isn't it as dangerous to be hit by the gun as by the bullet?

Answer   It is the product mv, which is the same for both the bullet and the gun. The bullet has a large velocity and a small mass, while the gun has a small velocity and a large mass. Furthermore, the bullet carries much more kinetic energy than the gun.

Q8.19   Does the center of mass of a rocket in free space accelerate? Explain. Can the speed of a rocket exceed the exhaust speed of the fuel? Explain.

Answer   The center of mass of a rocket, plus its exhaust does not accelerate: momentum must be conserved. However, if you consider the "rocket" to be the mechanical system plus the unexpended fuel, then it becomes obvious that the center of mass of that "rocket" does accelerate.

　　　　Basically, a rocket is no more complicated than two rubber balls that are pressed together, and released. One springs in one direction with one velocity; the other springs in the other direction with a velocity equal to

$$v_2 = v_1 \left(\frac{m_1}{m_2} \right)$$

In a similar manner, the mass of the rocket times its speed must equal the mass of all parts of the exhaust, times their speed. After a sufficient quantity of fuel has been exhausted the ratio $\dfrac{m_1}{m_2}$ will be greater than 1; then the speed of the rocket **can** exceed the speed of the exhaust.

SOLUTIONS TO SELECTED PROBLEMS

P8.7   An estimated force-time curve for a baseball struck by a bat is shown in Figure P8.7.

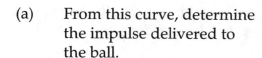 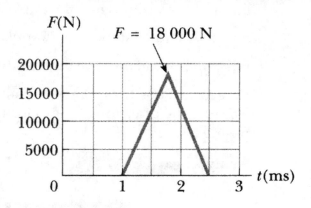

FIG. P8.7

(a)   From this curve, determine the impulse delivered to the ball.

(b)   From this curve, determine the average force exerted on the ball.

(c)   From this curve, determine the peak force exerted on the ball.

Solution The impulse delivered to the ball is equal to the area under the *F-t* graph. Thus,

(a) $I = \left(\dfrac{0 + 18000 \text{ N}}{2}\right)\left(2.5 \times 10^{-3} \text{ s} - 1.0 \times 10^{-3} \text{ s}\right) = 13.5 \text{ N} \cdot \text{s}$ ◊

(b) $F_{av} = \dfrac{\int F\,dt}{\Delta t} = \dfrac{13.5 \text{ N} \cdot \text{s}}{(2.5 - 1.0)10^{-3} \text{ s}} = 9.00 \text{ kN}$ ◊

(c)   From the graph, $F_{max} = 18\,000 \text{ N}$ ◊

P8.9   A 3.00-kg steel ball strikes a wall with a speed of 10.0 m/s at an angle of 60.0° with the surface. It bounces off with the same speed and angle (Fig. P8.9). If the ball is in contact with the wall for 0.200 s, what is the average force exerted on the ball by the wall?

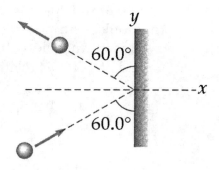

FIG. P8.9

continued on next page

Solution **Conceptualize:** If we think about the angle as a variable and consider the limiting cases, then the force should be zero when the angle is 0° (no contact between the ball and the wall). When the angle is 90° the force will be its maximum and can be found from the momentum-impulse equation, so that $F = \dfrac{\Delta p}{\Delta t} < 300$ N, and the force on the ball must be directed to the left.

Categorize: Use the momentum-impulse equation to find the force, and carefully consider the direction of the velocity vectors by defining up and to the right as positive.

Analyze: $\Delta \vec{p} = \vec{F} \Delta t$:

$$\Delta p_y = m\left(v_{fy} - v_{iy}\right) = m(v\cos 60.0° - v\cos 60.0°) = 0.$$

So the wall does not exert a force on the ball in the y direction.

$$\Delta p_x = m\left(v_{fx} - v_{ix}\right) = m(-v\sin 60.0° - v\sin 60.0°) = -2mv\sin 60.0°$$

$$\Delta p_x = -2(3.00 \text{ kg})(10.0 \text{ m/s})(0.866) = -52.0 \text{ kg} \cdot \text{m/s}$$

$$\vec{F}_{av} = \frac{\Delta \vec{p}}{\Delta t} = \frac{\Delta p_x \hat{i}}{\Delta t} = \frac{-52.0 \hat{i} \text{ kg} \cdot \text{m/s}}{0.200 \text{ s}} = -260 \hat{i} \text{ N} \qquad \lozenge$$

Finalize: The force is to the left and has a magnitude less than 300 N as expected.

P8.15 A 45.0-kg girl is standing on a plank that has a mass of 150 kg. The plank, originally at rest, is free to slide on a frozen lake, which is a flat, frictionless supporting surface. The girl begins to walk along the plank at a constant speed of 1.50 m/s relative to the plank.

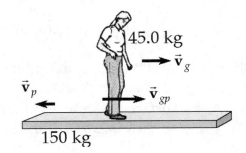

FIG. P8.15

(a) What is her speed relative to the ice surface?

(b) What is the speed of the plank relative to the ice surface?

continued on next page

Solution $\vec{\mathbf{v}}_g$ = velocity of the girl relative to the ice

$\vec{\mathbf{v}}_{gp}$ = velocity of the girl relative to the plank

$\vec{\mathbf{v}}_p$ = velocity of the plank relative to the ice

The girl and the plank exert forces on each other, but the ice isolates them from outside horizontal forces. Therefore, the net momentum is zero for the combined girl plus plank system.

$$0 = m_g \vec{\mathbf{v}}_g + m_p \vec{\mathbf{v}}_p.$$

Further, the relation among relative speeds can be written $\vec{\mathbf{v}}_g = \vec{\mathbf{v}}_{gp} + \vec{\mathbf{v}}_p$

$$\vec{\mathbf{v}}_g = 1.50\hat{\mathbf{i}} \text{ m/s} + \vec{\mathbf{v}}_p.$$

We substitute:

$$0 = (45.0 \text{ kg})(1.50\hat{\mathbf{i}} \text{ m/s} + \vec{\mathbf{v}}_p) + (150 \text{ kg})\vec{\mathbf{v}}_p$$

$$(195 \text{ kg})\vec{\mathbf{v}}_p = -(45.0 \text{ kg})(1.50\hat{\mathbf{i}} \text{ m/s})$$

(b) $\vec{\mathbf{v}}_p = -0.346\hat{\mathbf{i}} \text{ m/s}$ ◊

(a) $\vec{\mathbf{v}}_g = 1.50\hat{\mathbf{i}} - 0.346\hat{\mathbf{i}} \text{ m/s} = 1.15\hat{\mathbf{i}} \text{ m/s}$ ◊

P8.19   A neutron in a nuclear reactor makes an elastic head-on collision with the nucleus of a carbon atom initially at rest.

(a)   What fraction of the neutron's kinetic energy is transferred to the carbon nucleus?

(b)   Assume that the initial kinetic energy of the neutron is 1.60×10^{-13} J. Find its final kinetic energy and the kinetic energy of the carbon nucleus after the collision. (The mass of the carbon nucleus is about 12.0 times the mass of the neutron.)

continued on next page

Solution (a)   This is a perfectly elastic head-on collision, so we use Eq. 8.21:

$$v_{1i} - v_{2i} = -\left(v_{1f} - v_{2f}\right).$$

Let object 1 be the neutron, and object 2 be the carbon nucleus, with $m_2 = 12m_1$.

Since $v_{2i} = 0$, $v_{2f} = v_{1i} + v_{1f}$.

Now, by conservation of momentum, $m_1 v_{1i} + m_2 v_{2i} = m_1 v_{1f} + m_2 v_{2f}$

or $m_1 v_{1i} = m_1 v_{1f} + 12 m_1 v_{2f}$.

Substituting our velocity equation, $v_{1i} = v_{1f} + 12(v_{1i} + v_{1f})$.

We solve $-11 v_{1i} = 13 v_{1f}$: $v_{1f} = -\dfrac{11}{13} v_{1i}$

And $v_{2f} = v_{1i} - \dfrac{11}{13} v_{1i} = \dfrac{2}{13} v_{1i}$.

The neutron's original kinetic energy is $\dfrac{1}{2} m_1 v_{1i}^2$.

The carbon's final kinetic energy is

$$\frac{1}{2} m_2 v_{2f}^2 = \frac{1}{2}(12 m_1)\left(\frac{2}{13}\right)^2 v_{1i}^2 = \left(\frac{48}{169}\right)\left(\frac{1}{2}\right) m_1 v_{1i}^2.$$

So, $\dfrac{48}{169} = 0.284$ or 28.4% of the total energy is transferred.   ◊

(b)   For the carbon nucleus, $K_{2f} = (0.284)(1.60 \times 10^{-13}\ \text{J}) = 45.4\ \text{fJ}$.   ◊

The collision is perfectly elastic, so the neutron retains the rest of the energy,

$$K_{1f} = (1.60 - 0.454) \times 10^{-13}\ \text{J} = 1.15 \times 10^{-13}\ \text{J} = 115\ \text{fJ}.$$ ◊

P8.21   A 12.0-g wad of sticky clay is hurled horizontally at a 100-g wooden block initially at rest on a horizontal surface. The clay sticks to the block. After impact, the block slides 7.50 m before coming to rest. If the coefficient of friction between the block and the surface is 0.650, what was the speed of the clay immediately before impact?

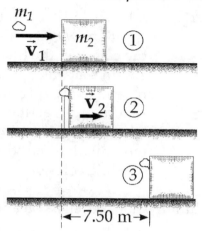

FIG. P8.21

Solution We use the momentum version of the isolated system model to analyze the collision, and the energy version of the isolated system model to analyze the subsequent sliding process. The collision, for which figures (1) and (2) are before and after pictures, is **totally inelastic**, and momentum is conserved for the system of clay and block:

$$m_1 v_1 = (m_1 + m_2)v_2.$$

In the sliding process occurring between figures (2) and (3), the original kinetic energy of the surface, block, and clay is equal to the increase in internal energy of the system due to friction:

$$\frac{1}{2}(m_1 + m_2)v_2{}^2 = f_f L$$

$$\frac{1}{2}(m_1 + m_2)v_2{}^2 = \mu(m_1 + m_2)gL$$

Solving for v_2: $v_2 = \sqrt{2\mu L g} = \sqrt{2(0.650)(7.50 \text{ m})(9.80 \text{ m/s}^2)} = 9.77$ m/s. From the momentum conservation equation,

$$v_1 = \left(\frac{m_1 + m_2}{m_1}\right)v_2 = \left(\frac{112 \text{ g}}{12.0 \text{ g}}\right)9.77 \text{ m/s} = 91.2 \text{ m/s} .$$

◊

P8.27   A billiard ball moving at 5.00 m/s strikes a stationary ball of the same mass. After the collision, the first ball moves at 4.33 m/s, at an angle of 30.0° with respect to the original line of motion. Assuming an elastic collision (and ignoring friction and rotational motion), find the struck ball's velocity after the collision.

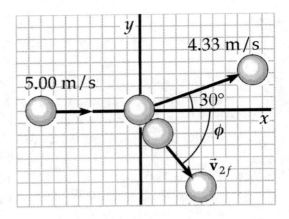

FIG. P8.27

Solution   Call each mass m, and call $\vec{v}_{2f}$ the velocity of the second ball after the collision, as in the figure. Then apply conservation of momentum to the two-ball system. In the x direction,

$$m(5.00 \text{ m/s}) = m(4.33 \text{ m/s})\cos 30° + mv_{2fx}$$

$$v_{2fx} = 1.25 \text{ m/s}$$

In the y direction,

$$0 = m(4.33 \text{ m/s})\sin 30° + mv_{2fy}$$

$$v_{2fy} = -2.17 \text{ m/s}$$

$$\vec{v}_{2f} = 1.25\hat{i} - 2.17\hat{j} \qquad \text{(or 2.50 m/s at 60.0°)} \qquad \lozenge$$

We did not have to use the fact that the collision is elastic.

P8.29   An object of mass 3.00 kg, moving with an initial velocity of $5.00\hat{\mathbf{i}}$ m/s, collides with and sticks to an object of mass 2.00 kg with an initial velocity of $-3.00\hat{\mathbf{j}}$ m/s. Find the final velocity of the composite object.

Solution   Momentum of the two-object system is conserved, with both objects having the same final velocity:

$$m_1\vec{\mathbf{v}}_{1i} + m_2\vec{\mathbf{v}}_{2i} = m_1\vec{\mathbf{v}}_{1f} + m_2\vec{\mathbf{v}}_{2f}$$

$$(3.00 \text{ kg})\left(5.00\hat{\mathbf{i}} \text{ m/s}\right) + (2.00 \text{ kg})\left(-3.00\hat{\mathbf{j}} \text{ m/s}\right) = (3.00 \text{ kg} + 2.00 \text{ kg})\vec{\mathbf{v}}_f$$

$$\vec{\mathbf{v}}_f = \frac{15.0\hat{\mathbf{i}} - 6.00\hat{\mathbf{j}}}{5.00} \text{ m/s} = \left(3.00\hat{\mathbf{i}} - 1.20\hat{\mathbf{j}}\right) \text{ m/s} \qquad \lozenge$$

Related Calculation: Compute the kinetic energy of the system both before and after the collision; show that kinetic energy is not conserved.

$$K_{1i} + K_{2i} = \frac{1}{2}(3.00 \text{ kg})(5.00 \text{ m/s})^2 + \frac{1}{2}(2.00 \text{ kg})(3.00 \text{ m/s})^2$$

$$K_{1i} + K_{2i} = 46.5 \text{ J}$$

$$K_{1f} + K_{2f} = \frac{1}{2}(5.00 \text{ kg})\left[(3.00 \text{ m/s})^2 + (1.20 \text{ m/s})^2\right] = 26.1 \text{ J} \qquad \lozenge$$

P8.31   An unstable atomic nucleus of mass 17.0×10^{-27} kg initially at rest disintegrates into three particles. One of the particles, of mass 5.00×10^{-27} kg, moves along the y axis with a velocity of 6.00×10^6 m/s. Another particle, of mass 8.40×10^{-27} kg, moves along the x axis with a speed of 4.00×10^6 m/s.

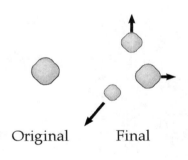

Original   Final

FIG. P8.31

(a)   Find the velocity of the third particle

(b)   Find the total kinetic energy increase in the process.

continued on next page

Solution (a)   With three particles, the total final momentum of the system is

$$m_1\vec{v}_{1f} + m_2\vec{v}_{2f} + m_3\vec{v}_{3f}$$

and it must be zero to equal the original momentum.

The mass of the third particle is $\qquad m_3 = (17.0 - 5.00 - 8.40) \times 10^{-27}$ kg

or $\qquad\qquad\qquad\qquad\qquad\qquad m_3 = 3.60 \times 10^{-27}$ kg.

The total momentum is zero: $\qquad m_1\vec{v}_{1f} + m_2\vec{v}_{2f} + m_3\vec{v}_{3f} = 0$.

Solving for $\vec{v}_{3f}$, $\qquad\qquad\qquad\qquad \vec{v}_{3f} = -\dfrac{m_1\vec{v}_{1f} + m_2\vec{v}_{2f}}{m_3}$.

$$\vec{v}_{3f} = -\frac{(3.00\hat{j} + 3.36\hat{i}) \times 10^{-20} \text{ kg} \cdot \text{m/s}}{3.60 \times 10^{-27} \text{ kg}} = \left(-9.33\hat{i} - 8.33\hat{j}\right) \text{M m/s} \qquad \Diamond$$

(b)   The original kinetic energy of the system is zero. The final kinetic energy is $K = K_{1f} + K_{2f} + K_{3f}$.

$$K_{1f} = \frac{1}{2}\left(5.00 \times 10^{-27} \text{ kg}\right)\left(6.00 \times 10^{6} \text{ m/s}\right)^2 = 9.00 \times 10^{-14} \text{ J}$$

$$K_{2f} = \frac{1}{2}\left(8.40 \times 10^{-27} \text{ kg}\right)\left(4.00 \times 10^{6} \text{ m/s}\right)^2 = 6.72 \times 10^{-14} \text{ J}$$

$$K_{3f} = \frac{1}{2}\left(3.60 \times 10^{-27} \text{ kg}\right)\left(\left(-9.33 \times 10^{6} \text{ m/s}\right)^2 + \left(-8.33 \times 10^{6} \text{ m/s}\right)^2\right)$$

$$= 28.2 \times 10^{-14} \text{ J}$$

and $K = 9.00 \times 10^{-14}$ J $+ 6.72 \times 10^{-14}$ J $+ 28.2 \times 10^{-14}$ J

$\qquad K = 4.39 \times 10^{-13}$ J $= 439$ fJ. $\qquad\qquad\qquad\qquad\qquad\qquad\qquad\qquad \Diamond$

P8.33 A uniform piece of sheet steel is shaped as in Figure P8.33. Compute the x and y coordinates of the center of mass of the piece.

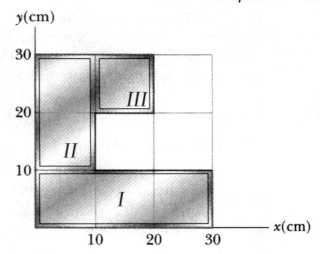

FIG. P8.33

Solution **Conceptualize:** By inspection, it appears that the center of mass is located at about

$$\left(12\hat{\mathbf{i}} + 13\hat{\mathbf{j}}\right) \text{ cm}$$

Categorize: Think of the sheet as composed of three sections, and consider the mass of each section to be at the geometric center of that section. Define the mass per unit area to be σ, and number the rectangles as shown. We can then calculate the mass and identify center of mass of each section.

Analyze: $m_I = (30.0 \text{ cm})(10.0 \text{ cm})\sigma \quad CM_I = (15.0 \text{ cm}, 5.00 \text{ cm})$

$$m_{II} = (10.0 \text{ cm})(20.0 \text{ cm})\sigma \quad CM_{II} = (5.0 \text{ cm}, 20.0 \text{ cm})$$

$$m_{III} = (10.0 \text{ cm})(10.0 \text{ cm})\sigma \quad CM_{III} = (15.0 \text{ cm}, 25.0 \text{ cm})$$

The overall center of mass is at a point defined by the vector equation $\vec{\mathbf{r}}_{CM} \equiv \dfrac{\Sigma m_i \vec{\mathbf{r}}_i}{\Sigma m_i}$. Substituting the appropriate values, $\vec{\mathbf{r}}_{CM}$ is calculated to be:

$$\vec{\mathbf{r}}_{CM} = \frac{\sigma\left[(300)\left(15.0\hat{\mathbf{i}} + 5.0\hat{\mathbf{j}}\right) + (200)\left(5.0\hat{\mathbf{i}} + 20.0\hat{\mathbf{j}}\right) + (100)\left(15.0\hat{\mathbf{i}} + 25.0\hat{\mathbf{j}}\right) \text{ cm}^3\right]}{\sigma\left(300 \text{ cm}^2 + 200 \text{ cm}^2 + 100 \text{ cm}^2\right)}$$

$$\vec{\mathbf{r}}_{CM} = \frac{\left(45.0\hat{\mathbf{i}} + 15.0\hat{\mathbf{j}} + 10.0\hat{\mathbf{i}} + 40.0\hat{\mathbf{j}} + 15.0\hat{\mathbf{i}} + 25.0\hat{\mathbf{j}}\right)}{6.00} \text{ cm}$$

$$\vec{\mathbf{r}}_{CM} = \left(11.7\hat{\mathbf{i}} + 13.3\hat{\mathbf{j}}\right) \text{ cm} \qquad \Diamond$$

continued on next page

Finalize: The coordinates are close to our eyeball estimate. In solving this problem, we could have chosen to divide the original shape some other way, but the answer would be the same. This problem also shows that the center of mass can lie outside the boundary of the object.

P8.37 A 2.00-kg particle has a velocity of $\left(2.00\hat{\mathbf{i}} - 3.00\hat{\mathbf{j}}\right)$ m/s, and a 3.00-kg particle has a velocity of $\left(1.00\hat{\mathbf{i}} + 6.00\hat{\mathbf{j}}\right)$ m/s.

(a) Find the velocity of the center of mass

(b) Find the total momentum of the system of two particles.

Solution Use $\vec{\mathbf{v}}_{CM} = \dfrac{m_1\vec{\mathbf{v}}_1 + m_2\vec{\mathbf{v}}_2}{m_1 + m_2}$ and $\vec{\mathbf{p}}_{CM} = (m_1 + m_2)\vec{\mathbf{v}}_{CM}$:

(a) $\vec{\mathbf{v}}_{CM} = \dfrac{(2.00 \text{ kg})\left[\left(2.00\hat{\mathbf{i}} - 3.00\hat{\mathbf{j}}\right) \text{ m/s}\right] + (3.00 \text{ kg})\left[\left(1.00\hat{\mathbf{i}} + 6.00\hat{\mathbf{j}}\right) \text{ m/s}\right]}{(2.00 \text{ kg} + 3.00 \text{ kg})}$

$\vec{\mathbf{v}}_{CM} = \left(1.40\hat{\mathbf{i}} + 2.40\hat{\mathbf{j}}\right)$ m/s   ◊

(b) $\vec{\mathbf{p}}_{CM} = (2.00 \text{ kg} + 3.00 \text{ kg})\left[\left(1.40\hat{\mathbf{i}} + 2.40\hat{\mathbf{j}}\right) \text{ m/s}\right] = \left(7.00\hat{\mathbf{i}} + 12.0\hat{\mathbf{j}}\right) \text{ kg} \cdot \text{m/s}$ ◊

P8.39 Romeo (77.0 kg) entertains Juliet (55.0 kg) by playing his guitar from the rear of their boat at rest in still water, 2.70 m away from Juliet in the front of the boat. After the serenade, Juliet carefully moves to the rear of the boat (away from shore) to plant a kiss on Romeo's cheek. How far does the 80.0-kg boat move toward the shore it is facing?

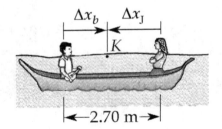

FIG. P8.39

Solution No outside forces act on the boat-plus-lovers system, so its momentum is conserved at zero and its center of mass stays fixed: $x_{CM,i} = x_{CM,f}$.

Define K to be the point where they kiss, and Δx_J and Δx_b as shown in the figure. Since Romeo moves with the boat (and thus $\Delta x_{Romeo} = \Delta x_b$), let m_b be the combined mass of Romeo and the boat. Then, $m_J \Delta x_J + m_b \Delta x_b = 0$.

continued on next page

Choosing the *x* axis to point away from the shore,

$$(55.0 \text{ kg})\Delta x_J + (77.0 \text{ kg} + 80.0 \text{ kg})\Delta x_b = 0 \qquad \text{and} \qquad \Delta x_J = -2.85\Delta x_b.$$

As Juliet moves away from shore, the boat and Romeo glide toward the shore until the original 2.70 m gap between them is closed:

$$\Delta x_J - \Delta x_b = 2.70 \text{ m}.$$

Substituting, we find $\Delta x_b = -0.700$ m, or 0.700 m towards the shore.   ◊

P8.41   The first stage of a Saturn V space vehicle consumed fuel and oxidizer at the rate of 1.50×10^4 kg/s, with an exhaust speed of 2.60×10^3 m/s.

(a)   Calculate the thrust produced by these engines.

(b)   Find the acceleration of the vehicle just as it lifted off the launch pad, if the vehicle's initial mass is 3.00×10^6 kg. You must include the gravitational force to solve part (b).

Solution   **Conceptualize:** The thrust must be at least equal to the weight of the rocket (≈ 30 MN); otherwise the launch would not have been successful! However, since Saturn V rockets accelerate rather slowly compared to the acceleration of falling objects, the thrust should be less than about twice the rocket's weight, so that $0 < a < g$.

Categorize: Use Newton's second law to find the force and acceleration from the changing momentum.

Analyze:

(a)   The thrust, *F*, is equal to the time rate of change of momentum as fuel is exhausted from the rocket.

$$F = \frac{dp}{dt} = \frac{d}{dt}(mv_e).$$

Since v_e is constant exhaust velocity, $F = v_e\left(\dfrac{dm}{dt}\right)$ where

$\dfrac{dm}{dt} = 1.50 \times 10^4$ kg/s and $v_e = 2.60 \times 10^3$ m/s.

$$F = \left(2.60 \times 10^3 \text{ m/s}\right)\left(1.50 \times 10^4 \text{ kg/s}\right) = 39.0 \text{ MN} \qquad ◊$$

continued on next page

(b)   Applying $\sum F = ma$:

$$\left(3.90 \times 10^7 \text{ N}\right) - \left(3.00 \times 10^6 \text{ kg}\right)\left(9.80 \text{ m/s}^2\right) = \left(3.00 \times 10^6 \text{ kg}\right)a$$

$$a = \frac{\left(3.90 \times 10^7 \text{ N}\right) - \left(29.4 \times 10^6 \text{ N}\right)}{3.00 \times 10^6 \text{ kg}} = 3.20 \text{ m/s}^2 \text{ up} \qquad \lozenge$$

Finalize: As expected, the thrust is slightly greater than the weight of the rocket, and the acceleration is about 0.3*g*, so the answers appear to be reasonable. This kind of rocket science is not so complicated after all!

Relativity

Chapter 9

NOTES FROM SELECTED CHAPTER SECTIONS

Section 9.1   The Principle of Newtonian Relativity

According to the principle of Newtonian relativity, the laws of mechanics are the same in all inertial frames of reference. *Inertial frames of reference are those coordinate systems at rest with respect to one another or which move at constant velocity with respect to one another.*

Section 9.2   The Michelson-Morley Experiment

This experiment was designed to detect the velocity of the Earth with respect to the hypothetical luminiferous ether. The outcome of the experiment was negative, contradicting the ether hypothesis.

Section 9.3   Einstein's Principle of Relativity

Einstein's special theory of relativity is based on two postulates:

- **The principle of relativity**: The laws of physics are the same in all inertial reference systems.

- **The constancy of the speed of light**: The speed of light in vacuum is always measured to be 3.00×10^8 m/s, and the value is independent of the motion of the observer or of the motion of the source of light.

The second postulate is consistent with the negative results of the Michelson-Morley experiment, which failed to detect the presence of an ether and suggested that the speed of light is the same in all inertial frames.

Section 9.4   Consequences of Special Relativity

- **Simultaneity**: Two events simultaneous in one reference frame are, in general, not simultaneous in another frame moving with respect to the first.

- **Time dilation**: According to a stationary observer, a moving clock runs slower than an identical clock at rest. The **proper time** is always the time measured by a clock at rest in the frame of reference of the measurement.

- **Length contraction**: The distance between any two points in space is measured by an observer to be contracted along the direction of the velocity of the observer relative to the points. The **proper length** of an object is defined as the length of the object measured in the reference frame in which the object is at rest. *The contraction occurs only along the direction of motion.*

Section 9.6   Relativistic Momentum and the Relativistic Form of Newton's Laws

To account for relativistic effects, it is necessary to modify the definition of momentum to satisfy the following conditions:

- The relativistic momentum must be conserved in all collisions.

- The relativistic momentum must approach the classical value, *mv*, as the quantity $\left(\dfrac{v}{c}\right)$ approaches zero.

EQUATIONS AND CONCEPTS

The **Lorentz coordinate transformation equations** allow transformation between reference frames for particles moving at speeds approaching the speed of light. *These equations represent the transformation between any two inertial frames in relative motion with velocity v in the x direction.*

$$\left.\begin{array}{l} t' = \gamma\left(t - \dfrac{v}{c^2}x\right) \\[2mm] x' = \gamma(x - vt) \end{array}\right\} \gamma \equiv \dfrac{1}{\sqrt{1 - \dfrac{v^2}{c^2}}} \qquad (9.8)$$

$$y' = y$$
$$z' = z$$

The **Lorentz velocity transformation** equations relate the observed velocity u' in the moving (S') frame to the measured velocity u in the S frame; v is the velocity of the S' frame relative to the S frame.

$$u'_x = \frac{u_x - v}{1 - \frac{u_x v}{c^2}} \tag{9.11}$$

$$u'_y = \frac{u_y}{\gamma\left(1 - \frac{u_x v}{c^2}\right)}$$

$$u'_z = \frac{u_z}{\gamma\left(1 - \frac{u_x v}{c^2}\right)} \tag{9.12}$$

Time dilation: *A clock moving with respect to an observer appears to run slower than an identical clock at rest with respect to the observer.* The **proper time** is the time measured by the observer in the reference frame of the clock; i.e. time interval between two events as measured by an observer for whom the events occur at the same point in space.

$$\Delta t = \frac{\Delta t_p}{\sqrt{1 - \frac{v^2}{c^2}}} = \gamma \Delta t_p \tag{9.6}$$

Δt_p = proper time

Length contraction: *The distance between two points in space as measured by an observer moving with respect to the points is contracted along the direction of relative motion.* The **proper length** of an object is defined as the distance in space between the endpoints of the object measured by someone who is at rest relative to the object.

$$L = L_p \left(1 - \frac{v^2}{c^2}\right)^{1/2} \tag{9.7}$$

ΔL_p = proper length

Simultaneity: *Two events that are simultaneous in one reference frame are not simultaneous in another reference frame that is in motion relative to the first.*

The **relativistic momentum** of a particle of mass m moving with speed $\vec{u}$ satisfies the conditions:

$$\vec{p} \equiv \frac{m\vec{u}}{\sqrt{1 - \frac{u^2}{c^2}}} \tag{9.14}$$

momentum is conserved in all collisions

$$\vec{p} = \gamma m \vec{u} \tag{9.15}$$

$\vec{p}$ approaches the classical expression as u approaches 0. That is, as $\vec{u} \Rightarrow 0$, $\vec{p} \Rightarrow m\vec{u}$.

The **relativistic kinetic energy** of a particle of mass m moving with speed u, is given by Equation 9.18. The term mc^2 is called the rest energy, E_R.

$$K = \frac{mc^2}{\sqrt{1 - \frac{u^2}{c^2}}} - mc^2 = (\gamma - 1)mc^2 \qquad (9.18)$$

The **total energy** E of a particle is the sum of the kinetic energy and the rest energy. *The relation* $E_R = mc^2$ *shows that mass is a manifestation of energy.*

$$E = \gamma mc^2 = K + mc^2 = K + E_R \qquad (9.20)$$

$$E = \frac{mc^2}{\sqrt{1 - \frac{u^2}{c^2}}} \qquad (9.21)$$

The **energy and momentum** of a relativistic particle are related as in Equation 9.22. *This expression is useful when the momentum and energy are known (rather than the speed).*

$$E^2 = p^2 c^2 + \left(mc^2\right)^2 \qquad (9.22)$$

For **zero-mass particles** (e.g. photons), Equation 9.23 is an exact expression relating energy and momentum. *These zero-mass particles always travel at the speed of light.*

$$E = pc \qquad (9.23)$$

The **electron volt** (eV) is a convenient energy unit to use to express the energies of electrons and other subatomic particles.

$$1 \text{ eV} = 1.60 \times 10^{-19} \text{ J}$$

REVIEW CHECKLIST

✓ State Einstein's two postulates of the special theory of relativity.

✓ Understand the Michelson-Morley experiment—its objectives, results, and the significance of its outcome.

✓ Understand the idea of simultaneity, and the fact that simultaneity is not an absolute concept. That is, two events simultaneous in one reference frame are not simultaneous when viewed from a second frame moving with respect to the first.

✓ Make calculations using the equations for time dilation, relativistic momentum, and length contraction.

✓ State the correct relativistic expressions for the momentum, kinetic energy, and total energy of a particle. Make calculations using these equations.

ANSWERS TO SELECTED QUESTIONS

Q9.5 Explain why it is necessary, when defining the length of a rod, to specify that the positions of the ends of the rod are to be measured simultaneously.

Answer Suppose a railroad train is moving past you. One way to measure its length is this: You mark the tracks at the front of the moving engine at 9:00:00 AM, while your assistant marks the tracks at the back of the caboose at the same time. Then you find the distance between the marks on the tracks with a tape measure. You and your assistant must make the marks simultaneously (in your reference frame), for otherwise the motion of the train would make its length different from the distance between marks.

Q9.7 List some ways our day-to-day lives would change if the speed of light were only 50 m/s.

Answer For a wonderful fictional exploration of this question, get a "Mr. Tompkins" book by George Gamow. All of the relativity effects would be obvious in our lives. Time dilation and length contraction would both occur. Driving home in a hurry, you would push on the gas pedal not to increase your speed very much, but to make the blocks shorter. Big Doppler shifts in wave frequencies would make red lights look green as you approached, and make car horns and radios useless. High-speed transportation would be both very expensive, requiring huge fuel purchases, as well as dangerous, since a speeding car could knock down a building. When you got home, hungry for lunch, you would find that you had missed dinner; there would be a five-day delay in transit when you watched the Olympics in Australia on live TV. Finally, we would not be able to see the Milky Way, since the fireball of the Big Bang would surround us at the distance of Rigel, or Deneb.

Q9.9 Give a physical argument that shows that it is impossible to accelerate an object of mass m to the speed of light, even with a continuous force acting on it.

Answer As an object approaches the speed of light, its energy approaches infinity. Hence, it would take an infinite amount of work to accelerate the object to the speed of light under the action of a continuous force or it would take an infinitely large force.

SOLUTIONS TO SELECTED PROBLEMS

P9.1   In a laboratory frame of reference, an observer notes that Newton's second law is valid. Show that it is also valid for an observer moving at a constant speed, small compared to the speed of light, relative to the laboratory frame.

Solution   The first observer watches some object accelerate along the x direction under applied forces. Call the instantaneous velocity of the object u_x. The second observer moves along the x axis at constant speed v relative to the first, and measures the object to have velocity

$$u'_x = u_x - v.$$

The acceleration is $\dfrac{du'_x}{dt} = \dfrac{du_x}{dt} - 0.$

This is the same as that measured by the first observer. In this nonrelativistic case, they measure also the same mass and forces; so the second observer also confirms that $\sum F = ma$.   ◊

P9.9   An atomic clock moves at 1 000 km/h for 1.00 h as measured by an identical clock on the Earth. How many nanoseconds slow will the moving clock be compared with the Earth clock at the end of the one-hour interval?

Solution   This problem is slightly more difficult than most, for the simple reason that your calculator probably cannot hold enough decimal places to yield an accurate answer. However, we can bypass the difficulty by noting the approximation:

$\sqrt{1 - \dfrac{v^2}{c^2}} \approx 1 - \dfrac{v^2}{2c^2}$. Squaring both sides will show that when $\dfrac{v}{c}$ is small, these two

expressions are equivalent. Evaluating $\dfrac{v}{c}$,

$$\frac{v}{c} = \left(\frac{1000 \times 10^3 \text{ m/h}}{3.00 \times 10^8 \text{ m/s}} \right)\left(\frac{1 \text{ h}}{3600 \text{ s}} \right) = 9.26 \times 10^{-7}.$$

From Equation 9.6, $\Delta t = \gamma \Delta t_p = \dfrac{\Delta t_p}{\sqrt{1 - v^2/c^2}}.$

Rearranging, our approximation yields $\Delta t_p = \left(\sqrt{1 - \dfrac{v^2}{c^2}} \right)\Delta t \approx \left(1 - \dfrac{v^2}{2c^2} \right)\Delta t$

continued on next page

and

$$\Delta t - \Delta t_p = \frac{v^2}{2c^2}\Delta t.$$

Substituting,

$$\Delta t - \Delta t_p = \frac{\left(9.26 \times 10^{-7}\right)^2}{2}(3\ 600\ \text{s}).$$

Thus the time lag of the moving clock is $\Delta t - \Delta t_p = 1.54 \times 10^{-9}$ s $= 1.54$ ns.

P9.11 A spaceship with a proper length of 300 m takes 0.750 μs to pass an Earth observer. Determine its speed as measured by the Earth observer.

Solution **Conceptualize:** We should first determine if the spaceship is traveling at a relativistic speed: classically, $v = \dfrac{300\ \text{m}}{0.750\ \mu s} = 4.00 \times 10^8$ m/s, which is faster than the speed of light (impossible)! Quite clearly, the relativistic correction must be used to find the correct speed of the spaceship, which we can guess will be close to the speed of light.

Categorize: We can use the contracted length equation to find the speed of the spaceship in terms of the proper length and the time. The time of 0.750 μs is the **proper time** measured by the Earth observer, because it is the time interval between two events that she sees as happening at the same point in space. The two events are the passage of the front end of the spaceship over her stopwatch, and the passage of the back end of the ship.

Analyze: $L = \dfrac{L_p}{\gamma}$, with $L = v\Delta t$: $\qquad v\Delta t = L_p\left(1 - \dfrac{v^2}{c^2}\right)^{1/2}.$

Squaring both sides,

$$v^2 \Delta t^2 = L_p^{\ 2}\left(1 - \frac{v^2}{c^2}\right)$$

$$v^2 c^2 = \frac{L_p^{\ 2} c^2}{\Delta t^2} - \frac{v^2 L_p^{\ 2}}{\Delta t^2}$$

Solving for the speed,

$$v = \frac{c L_p/\Delta t}{\sqrt{c^2 + L_p^{\ 2}/\Delta t^2}} = \frac{\left(3.00 \times 10^8\ \text{m/s}\right)(300\ \text{m})/\left(0.750 \times 10^{-6}\ \text{s}\right)}{\sqrt{\left(3.00 \times 10^8\ \text{m/s}\right)^2 + (300\ \text{m})^2/\left(0.750 \times 10^{-6}\ \text{s}\right)^2}}.$$

So $v = 2.40 \times 10^8$ m/s. $\qquad\qquad\qquad\Diamond$

continued on next page

Finalize: The spaceship is traveling at 0.8c. We can also verify that the general equation for the speed reduces to the classical relation $v = \dfrac{L_p}{\Delta t}$ when the time is relatively large.

P9.21 Two jets of material from the center of a radio galaxy are ejected in opposite directions. Both jets move at 0.750c relative to the galaxy. Determine the speed of one jet relative to the other.

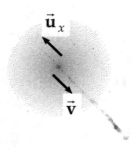

Solution Take the galaxy as the unmoving frame. Arbitrarily define the jet moving upwards to be the object, and the jet moving downwards to be the "moving" frame.

FIG. P9.21

$$u_x = 0.750c \qquad\qquad v = -0.750c$$

Thus $u'_x = \dfrac{u_x - v}{1 - u_x v / c^2} = \dfrac{0.750c - (-0.750c)}{1 - (0.750c)(-0.750c)/c^2} = \dfrac{1.50c}{1 + 0.750^2} = 0.960c.$ ◊

P9.27 An unstable particle at rest breaks into two fragments of unequal mass. The mass of one fragment is 2.50×10^{-28} kg, and that of the other is 1.67×10^{-27} kg. If the lighter fragment has a speed of 0.893c after the breakup, what is the speed of the heavier fragment?

Solution **Conceptualize:** The heavier fragment should have a speed less than that of the lighter piece since the momentum of the system must be conserved. However, due to the relativistic factor, the ratio of the speeds will not equal the simple ratio of the particle masses, which would give a speed of 0.134c for the heavier particle.

Categorize: Relativistic momentum of the system must be conserved. For the total momentum to be zero after the fission, as it was before, $\vec{p}_1 + \vec{p}_2 = 0$, where we will refer to the lighter particle with the subscript '1', and to the heavier particle with the subscript '2.'

Analyze:

$$\gamma_2 m_2 v_2 + \gamma_1 m_1 v_1 = 0: \qquad \gamma_2 m_2 v_2 + \left(\dfrac{2.50 \times 10^{-28} \text{ kg}}{\sqrt{1 - 0.893^2}} \right)(0.893c) = 0.$$

Rearranging, $\left(\dfrac{1.67 \times 10^{-27} \text{ kg}}{\sqrt{1 - v_2^2/c^2}} \right) \dfrac{v_2}{c} = -4.96 \times 10^{-28} \text{ kg}.$

continued on next page

Squaring both sides, $\left(2.79\times10^{-54}\ \text{kg}^2\right)\left(\dfrac{v_2}{c}\right)^2 = \left(2.46\times10^{-55}\ \text{kg}^2\right)\left(1-\dfrac{v_2^2}{c^2}\right)$

and $\qquad v_2 = -0.285c$.

We choose the negative sign only to mean that the two particles must move in **opposite** directions. The speed, then, is $|v_2| = 0.285c$. ◊

Finalize: The speed of the heavier particle is less than the lighter particle, as expected. We can also see that for this situation, the relativistic speed of the heavier particle is about twice as great as was predicted by a simple non-relativistic calculation.

P9.33　　A proton moves at $0.950c$.

　　(a)　　Calculate its rest energy.

　　(b)　　Calculate its total energy.

　　(c)　　Calculate its kinetic energy.

Solution　At $v = 0.950c$, $\gamma = \dfrac{1}{\sqrt{1-v^2/c^2}} = \dfrac{1}{\sqrt{1-0.950^2}} = 3.20$

　　(a)　　$E_R = mc^2 = (1.67\times10^{-27}\ \text{kg})(2.998\times10^8\ \text{m/s})^2 = 1.50\times10^{-10}\ \text{J} = 938\ \text{MeV}$ ◊

　　(b)　　$E = \gamma mc^2 = \gamma E_R = (3.20)(938\ \text{MeV}) = 3.00\ \text{GeV}$ ◊

　　(c)　　$K = E - E_R = 3.00\ \text{GeV} - 938\ \text{MeV} = 2.07\ \text{GeV}$ ◊

P9.37　　Show that the energy-momentum relationship $E^2 = p^2c^2 + \left(mc^2\right)^2$ follows from the expressions $E = \gamma mc^2$ and $p = \gamma mu$.

Solution　$E = \gamma mc^2$, $\qquad\qquad p = \gamma mu$.

Squaring both equations, $E^2 = \left(\gamma mc^2\right)^2$, $\qquad\qquad p^2 = \left(\gamma mu\right)^2$.

Multiplying the second equation by c^2, and subtracting it from the first,

$$E^2 - p^2c^2 = (\gamma mc^2)^2 - (\gamma mu)^2 c^2: \qquad E^2 - p^2c^2 = \gamma^2\left(\left(mc^2\right)\left(mc^2\right) - \left(mc^2\right)\left(mu^2\right)\right).$$

continued on next page

Extracting the $\left(mc^2\right)$ terms,

$$E^2 - p^2c^2 = \gamma^2\left(mc^2\right)^2\left(1-\frac{u^2}{c^2}\right)$$

and applying the definition of γ,

$$E^2 - p^2c^2 = \left(1-\frac{u^2}{c^2}\right)^{-1}\left(mc^2\right)^2\left(1-\frac{u^2}{c^2}\right).$$

Thus,

$$E^2 - p^2c^2 = \left(mc^2\right)^2. \qquad \diamondsuit$$

P9.39   A pion at rest $\left(m_\pi = 273m_e\right)$ decays to a muon $\left(m_\mu = 207m_e\right)$ and an antineutrino $\left(m_{\bar{\nu}} \approx 0\right)$. The reaction is written $\pi^- \rightarrow \mu^- + \bar{\nu}$. Find the kinetic energy of the muon and the energy of the antineutrino in electron volts. (*Suggestion:* Conserve both energy and momentum.)

Solution   We use, together, both the energy version and the momentum version of the isolated system model. By conservation of energy, $m_\pi c^2 = \gamma m_\mu c^2 + |p_\nu|c$. By conservation of momentum, $p_\nu = -p_\mu = -\gamma m_\mu v$. Substituting the second equation into the first, $m_\pi c^2 = \gamma m_\mu c^2 + \gamma m_\mu vc$. Simplified, this equation then reads

$m_\pi = m_\mu\left(\gamma + \frac{\gamma v}{c}\right)$. Substituting for the masses, $273m_e = (207m_e)\left(\gamma + \frac{\gamma v}{c}\right)$. Where

$m_e c^2 = 0.511$ MeV. Numerically, $\dfrac{273m_e}{207m_e} = \dfrac{1+v/c}{\sqrt{1-(v/c)^2}} = \sqrt{\dfrac{1+v/c}{1-v/c}}$. Solving for the

muon speed, $\dfrac{v}{c} = \dfrac{273^2 - 207^2}{273^2 + 207^2} = 0.270$. Therefore, $\gamma = \dfrac{1}{\sqrt{1-v^2/c^2}} = 1.038\,5$ and

$$K_\mu = (0.038\,5)(207 \times 0.511 \text{ MeV}) = 4.08 \text{ MeV} \qquad \diamondsuit$$
$$K_{\bar{\nu}} = (273 \times 0.511 \text{ MeV}) - (207 \times 0.511 \text{ MeV} + 4.08 \text{ MeV}) = 29.6 \text{ MeV} \qquad \diamondsuit$$

P9.41   The power output of the Sun is 3.85×10^{26} W. How much mass is converted to energy in the Sun each second?

Solution   From $E_R = mc^2$, we have for one second of solar operation

$$m = \frac{E_R}{c^2} = \frac{3.85 \times 10^{26} \text{ J}}{(3.00 \times 10^8 \text{ m/s})^2} = 4.28 \times 10^9 \text{ kg.} \qquad \diamondsuit$$

P9.49 The cosmic rays of highest energy are protons that have kinetic energy on the order of 10^{13} MeV.

(a) How long would it take a proton of this energy to travel across the Milky Way galaxy, having a diameter on the order of $\sim 10^5$ ly, as measured in the proton's frame?

(b) From the point of view of the proton, how many kilometers across is the galaxy?

Solution **Conceptualize:** We can guess that the energetic cosmic rays will be traveling close to the speed of light, so the time it takes a proton to traverse the Milky Way will be much less in the proton's frame than 10^5 years. The galaxy will also appear smaller to the high-speed protons than the galaxy's proper diameter of 10^5 light-years.

Categorize: The kinetic energy of the protons can be used to determine the relativistic γ-factor which can then be applied to the time dilation and length contraction equations to find the time and distance in the proton's frame of reference.

Analyze: The relativistic kinetic energy of a proton is $K = (\gamma - 1)mc^2 = 10^{13}$ MeV. Its rest energy is

$$mc^2 = \left(1.67 \times 10^{-27} \text{ kg}\right)\left(2.998 \times 10^8 \text{ m/s}\right)^2 \left(\frac{1 \text{ eV}}{1.60 \times 10^{-19} \text{ kg} \cdot \text{m}^2/\text{s}^2}\right) = 938 \text{ MeV}$$

So 10^{13} MeV $= (\gamma - 1)(938 \text{ MeV})$, and therefore $\gamma = 1.07 \times 10^{10}$.

The proton's speed in the galaxy's reference frame can be found from

$$\gamma = \frac{1}{\sqrt{1 - v^2/c^2}}: \qquad 1 - \frac{v^2}{c^2} = 8.80 \times 10^{-21} \text{ and}$$

$$v = c\sqrt{1 - 8.80 \times 10^{-21}} = \left(1 - 4.40 \times 10^{-21}\right)c \approx 3.00 \times 10^8 \text{ m/s}.$$

The proton's speed is nearly as large as the speed of light. In the galaxy frame, the traversal time is

$$\Delta t = \frac{x}{v} = \frac{10^5 \text{ light-years}}{c} = 10^5 \text{ years}.$$

continued on next page

(a)   This is dilated from the proper time measured in the proton's frame. The proper time is found from $\Delta t = \gamma \Delta t_p$:

$$\Delta t_p = \frac{\Delta t}{\gamma} = \frac{10^5 \text{ yr}}{1.07 \times 10^{10}} = 9.38 \times 10^{-6} \text{ years} = 296 \text{ s}$$

$\Delta t_p \sim$ a few hundred seconds   ◊

(b)   The proton sees the galaxy moving by at a speed nearly equal to c, passing in 296 s:

$$L = v\Delta t_p = \left(3.00 \times 10^8\right)(296 \text{ s}) = 8.88 \times 10^7 \text{ km} \sim 10^8 \text{ km} \qquad ◊$$

$$L = \left(8.88 \times 10^{10} \text{ m}\right)\left(\frac{1 \text{ ly}}{9.46 \times 10^{15} \text{ m}}\right) = 9.39 \times 10^{-6} \text{ ly} \sim 10^{-5} \text{ ly}$$

Finalize: The results agree with our predictions, although we may not have guessed that the protons would be traveling so close to the speed of light! The calculated results should be rounded to zero significant figures since we were given order of magnitude data. We should also note that the relative speed of motion v and the value of γ are the same in both the proton and galaxy reference frames.

P9.55   A supertrain (proper length 100 m) travels at a speed of 0.950c as it passes through a tunnel (proper length 50.0 m). As seen by a trackside observer, is the train ever completely within the tunnel? If so, with how much space to spare?

Solution The observer sees the proper length of the tunnel, 50.0 m, but sees the train Lorentz-contracted to length

$$L = L_p\sqrt{1 - \frac{v^2}{c^2}} = (100 \text{ m})\sqrt{1 - (0.950c)^2} = 31.2 \text{ m}$$

This is shorter than the tunnel by 18.8 m, so it is completely within the tunnel.   ◊

P9.57   A particle with electric charge q moves along a straight line in a uniform electric field $\vec{E}$ with a speed of u. The electric force exerted on the particle is $q\vec{E}$. If the motion and the electric field are both in the x direction.

(a)   Show that the acceleration of the charge q in the x direction is given by

$$a = \frac{du}{dt} = \frac{qE}{m}\left(1 - \frac{u^2}{c^2}\right)^{3/2}.$$

(b)   Discuss the significance of the dependence of the acceleration on the speed.

(c)   If the particle starts from rest at $x = 0$ at $t = 0$, how would you proceed to find the speed of the particle and its position at time t?

Solution The force on a charge in an electric field is given by $F = qE$. Further, at any speed, the **momentum** of the particle is given by $p = \gamma mu = \dfrac{mu}{\sqrt{1 - \frac{u^2}{c^2}}}$.

(a)   We use the momentum version of the nonisolated system model. Since

$$F = qE = \frac{dp}{dt}, \quad qE = \frac{d}{dt}\left(mu\left(1 - u^2/c^2\right)^{-1/2}\right)$$

$$qE = m\left(1 - \frac{u^2}{c^2}\right)^{-1/2}\frac{du}{dt} + \frac{1}{2}mu\left(1 - \frac{u^2}{c^2}\right)^{-3/2}\left(\frac{2u}{c^2}\right)\frac{du}{dt}$$

Simplifying, we find that $\dfrac{qE}{m} = \dfrac{du}{dt}\left(1 - \dfrac{u^2}{c^2}\right)^{-3/2}$ and $a = \dfrac{du}{dt} = \dfrac{qE}{m}\left(1 - \dfrac{u^2}{c^2}\right)^{3/2}$. ◊

(b)   As $u \to c$, we see that $a \to 0$. The particle thus never attains the speed of light. If $u \ll c$, $a \approx \dfrac{qE}{m}$, in agreement with the nonrelativistic account.

(c)   Taking the acceleration equation, isolating the velocity terms and integrating,

$$\int_0^u \left(1 - \frac{u^2}{c^2}\right)^{-3/2} du = \int_0^t \frac{qE}{m} dt: \qquad u = \frac{qEct}{\sqrt{m^2c^2 + q^2E^2t^2}} = \frac{dx}{dt} \qquad ◊$$

continued on next page

$$x = \int_0^x dx = qEc \int_0^t \frac{t\,dt}{\sqrt{m^2c^2 + q^2E^2t^2}} = \frac{c}{qE}\left(\sqrt{m^2c^2 + q^2E^2t^2} - mc\right) \qquad \Diamond$$

For early times, this expression predicts

$$x \approx \frac{c}{qE}\left[\left(m^2c^2\right)^{1/2}\left(1 + \frac{q^2E^2t^2}{2m^2c^2}\right) - mc\right] = \frac{qEt^2}{2m} = \frac{1}{2}\frac{F}{m}t^2,$$

in agreement with classical physics. A long time into the motion, the expression approaches $x \approx \frac{c}{qE}qEt = ct$, since the particle's speed approaches c.

Chapter 10

Rotational Motion

Section 10.1   Angular Position, Speed, and Angular Acceleration

Pure rotational motion refers to the motion of a rigid body about a fixed axis. In the case of **rotation about a fixed axis**, every particle on the rigid body has the same angular velocity and the same angular acceleration.

One **radian** (rad) is the angle subtended by an arc length equal to the radius of the arc.

The angular displacement (θ), angular velocity (ω), and angular acceleration (α) are analogous to linear displacement (x), linear velocity (v), and linear acceleration (a), respectively. The variables, θ, ω, and α, differ dimensionally from the variables x, v, and a, only by a length factor.

Section 10.2   Rotational Kinematics: the Rigid Body Under Constant Angular Acceleration

The **kinematic expressions** for rotational motion under constant angular acceleration are of the **same form** as those for linear motion under constant linear acceleration with the substitutions $x \rightarrow \theta$, $v \rightarrow \omega$, and $a \rightarrow \alpha$.

Section 10.3   Relationships Between Rotational and Translational Quantities

When a rigid body rotates about a fixed axis, every part of the body has the same angular velocity and the same angular acceleration. However, different parts of the body, in general, have different linear velocities and different linear accelerations.

Section 10.4 Rotational Kinetic Energy

From the definition of moment of inertia, we see that it has dimensions of ML^2 ($kg \cdot m^2$ in SI units). It plays the role of mass in all rotational equations. Although we shall commonly refer to the quantity $\frac{1}{2}I\omega^2$ as the rotational kinetic energy, it is not a new form of energy. It is ordinary kinetic energy. It is important to recognize the analogy between kinetic energy associated with linear motion, $\frac{1}{2}mv^2$, and rotational kinetic energy, $\frac{1}{2}I\omega^2$. The quantities I and ω in rotational motion are analogous to m and v in linear motion, respectively.

The **total kinetic energy** of a body in rolling motion is the sum of the rotational kinetic energy about the center of mass and the translational kinetic energy of the center of mass.

Section 10.5 Torque and the Vector Product

Torque is a physical quantity—the measure of the tendency of a force to cause rotation of a body about a specified axis. **Torque must be defined with respect to a specific axis of rotation.** Torque, which has the SI units of $N \cdot m$, must not be confused with force.

Section 10.6 The Rigid Body in Equilibrium

A rigid body is defined as one that does not deform under the application of external forces. There are two necessary conditions for equilibrium of a rigid body: (1) the resultant external force must be zero and (2) the resultant external torque must be zero about any axis. Two forces have an equivalent effect on a rigid body if, and only if, they have equal magnitudes and they have equal torques about any specified axis.

Section 10.8 Angular Momentum

The torque acting on a particle is equal to the time rate of change of its angular momentum.

Section 10.9 Conservation of Angular Momentum

The total angular momentum of a system is constant if the net external torque acting on the system is zero. The resultant torque acting about the center of mass of a body equals the time rate of change of angular momentum, regardless of the motion of the center of mass.

Section 10.11 Rolling of Rigid Bodies

Although the points on a rigid body rotating about a fixed axis may not experience the same force, linear acceleration, or linear velocity, every point on the body has the same angular acceleration and angular velocity at any instant. Therefore, at any instant the rigid body as a whole is characterized by specific values for angular acceleration, net torque, and angular velocity.

The **work-energy theorem in rotational motion** states that the net work done by external torques in rotating a rigid body about a fixed axis equals the change in the body's rotational kinetic energy.

The **total kinetic energy** of a body undergoing rolling motion is the sum of the rotational kinetic energy about the center of mass and the translational kinetic energy of the center of mass.

EQUATIONS AND CONCEPTS

The **arc length**, s, is the distance traveled by a particle as it moves along a circular path of radius r. *The radial line from the center of the circular path to the particle sweeps out an angle, θ.*

$$\theta = \frac{s}{r} \tag{10.1b}$$

The **angular displacement** θ is the ratio of two lengths (arc length to radius) and hence is a dimensionless quantity. However, it is common practice to refer to the angle as being in units of radians. In calculations, the relationship between radians and degrees is shown below the figure.

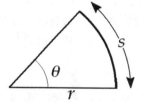

$$\theta^{rad} = \left(\frac{2\pi}{360°}\right)\theta^{deg} = \left(\frac{\pi}{180°}\right)\theta^{deg}$$

The **average angular speed** ω_{avg} of a particle or body rotating about a fixed axis equals the ratio of the angular displacement $\Delta\theta$ to the time interval Δt, where θ is measured in radians.

$$\omega_{avg} \equiv \frac{\theta_f - \theta_i}{t_f - t_i} = \frac{\Delta\theta}{\Delta t} \tag{10.2}$$

The **instantaneous angular speed** ω is defined as the limit of the average angular velocity as Δt approaches zero.

$$\omega \equiv \lim_{\Delta t \to 0} \frac{\Delta\theta}{\Delta t} = \frac{d\theta}{dt} \tag{10.3}$$

The **average angular acceleration** α_{avg} of a rotating body is defined as the ratio of the change in angular velocity to the time interval Δt.

$$\alpha_{avg} \equiv \frac{\omega_f - \omega_i}{t_f - t_i} = \frac{\Delta\omega}{\Delta t} \tag{10.4}$$

The **instantaneous angular acceleration** equals the limit of the average angular acceleration as Δt approaches zero.

$$\alpha \equiv \lim_{\Delta t \to 0} \frac{\Delta \omega}{\Delta t} = \frac{d\omega}{dt} \qquad (10.5)$$

The **equations of rotational kinematics** (10.6 through 10.9) apply when a particle or body rotates about a fixed axis with *constant angular acceleration.*

$$\omega_f = \omega_i + \alpha t \qquad (10.6)$$

$$\theta_f = \theta_i + \omega_i t + \frac{1}{2}\alpha t^2 \qquad (10.7)$$

See Table 10.1 of the text for a comparison of equations for rotational and translational motion.

$$\omega_f^2 = \omega_i^2 + 2\alpha(\theta_f - \theta_i) \qquad (10.8)$$

$$\theta_f = \theta_i + \frac{1}{2}(\omega_i + \omega_f)t \qquad (10.9)$$

The **tangential speed** of any point on a rigid body a distance r from a fixed axis of rotation is related to the angular speed via the radius.

$$v = r\omega \qquad (10.10)$$

The **tangential acceleration** of any point on a rotating rigid object is related to the angular acceleration through the relation $a_t = r\alpha$. *Note that every point on the object has the same ω and α, but the values of v and a_t depend on the radial distance from the axis.*

$$a_t = r\alpha \qquad (10.11)$$

The **centripetal acceleration** is directed toward the center of rotation. *The magnitude of $\bar{a}_r$ equals* a_c.

$$a_c = \frac{v^2}{r} = r\omega^2 \qquad (10.12)$$

The **total linear acceleration** has both radial and tangential components and a magnitude given by Equation 10.13.

$$a = \sqrt{a_t^2 + a_r^2} = r\sqrt{\alpha^2 + \omega^4} \qquad (10.13)$$

The **moment of inertia**, I is a measure of a system's resistance to a change in its angular speed.

$$I = \sum_i m_i r_i^2 \text{ (system of particles)} \qquad (10.14)$$

$$I = \int r^2 dm \text{ (continuous object)} \qquad (10.16)$$

The **torque** τ due to an applied force has a magnitude given by the product of the force and its moment arm d, where d equals the perpendicular distance from the rotation axis to the line of action of $\vec{F}$. Torque is a measure of the ability of a force to rotate a body about a specified axis. Note that the torque depends on the axis of rotation, which must be specified when τ is evaluated.

$$\pi \equiv rF\sin\phi = Fd \qquad (10.18)$$

$$d = \text{moment arm}$$

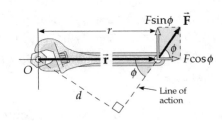

The **torque acting on a particle** whose vector position is $\vec{r}$ can be expressed as $\vec{r} \times \vec{F}$, where $\vec{F}$ is the external force acting on the particle. Torque also depends on the choice of the origin and has the SI unit of $N \cdot m$.

$$\vec{\tau} = \vec{r} \times \vec{F} \qquad (10.19)$$

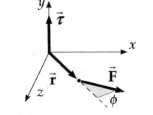

The **cross product of any two vectors** $\vec{A}$ and $\vec{B}$, is a vector $\vec{C}$ whose magnitude is given by $AB\sin\theta$ and whose direction is perpendicular to the plane formed by $\vec{A}$ and $\vec{B}$. The sense of $\vec{C}$ can be determined from the right-hand rule. See Section 10.5 of the text for properties of the unit vectors.

$$\vec{C} = \vec{A} \times \vec{B} \qquad (10.20)$$

$$C = \left|\vec{C}\right| \equiv AB\sin\theta \qquad (10.21)$$

The **first condition of equilibrium** corresponds to the condition of translational equilibrium. *When an object is at rest or is moving with constant velocity the resultant force on the object must be zero.*

$$\sum \vec{F} = 0 \qquad (10.24)$$

$$\sum F_x = 0 \quad \sum F_y = 0 \quad \sum F_z = 0$$

The **second condition of equilibrium** of a rigid body requires that the vector sum of the torques relative to any origin must be zero. *This is a statement of rotational equilibrium and requires that the angular acceleration about any axis be zero.*

$$\sum \vec{\tau} = 0 \qquad (10.25)$$

In the **case of coplanar forces** (i.e. all forces in the xy plane) three equations specify equilibrium. Two equations correspond to the first condition of equilibrium and the third comes from the second condition (the torque equation). *In this case, the torque vector lies along a line parallel to the z axis. All problems in this chapter fall into this category.*

$$\sum F_x = 0$$
$$\sum F_y = 0$$
$$\sum \tau_z = 0$$

(10.26)

The **rotational form of Newton's second law** states that the net torque acting on a rigid object rotating about a fixed axis equals the product of the moment of inertia about the axis of rotation and the angular acceleration relative to that axis.

$$\sum \tau = I\alpha$$

(10.27)

The **work-kinetic energy theorem for pure rotation** is given by Equation 10.29.

$$W = \frac{1}{2}I\omega_f^2 - \frac{1}{2}I\omega_i^2$$

(10.29)

The **angular momentum of a particle** whose linear momentum is $\vec{p}$ and whose vector position is $\vec{r}$ is defined as $\vec{L} = \vec{r} \times \vec{p}$. The SI unit of angular momentum is $kg \cdot m/s^2$. Note that both the magnitude and direction of $\vec{L}$ depend on the choice of origin.

$$\vec{L} \equiv \vec{r} \times \vec{p}$$

(10.32)

The **torque on a particle** equals the time rate of change of its angular momentum when the same origin is used to define angular momentum and torque. *This expression is the basic equation for treating rotating rigid bodies and rotating particles. Equation 10.36 is valid for any origin fixed in an inertial frame.*

$$\vec{\tau} = \frac{d\vec{L}}{dt}$$

(10.36)

The **net torque acting on a system of particles** equals the time rate of change of the total angular momentum of the system.

$$\sum \vec{\tau}_{ext} = \frac{d\vec{L}_{tot}}{dt}$$

(10.37)

The **magnitude of the angular momentum of a rigid body** in the form of a plane lamina rotating in the *x-y* plane about a **fixed axis** (the *z* axis) is given by the product $I\omega$, where I is the moment of inertia about the axis of rotation and ω is the angular speed.

$$L_z = I\omega$$

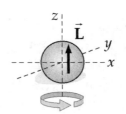

The **law of conservation of angular momentum** states that if the resultant external torque acting on a system is zero, the total angular momentum is constant. This follows from Equation 10.31.

$$\sum \vec{\tau}_{\text{ext}} = \frac{d\vec{L}_{\text{tot}}}{dt} = 0 \qquad (10.38)$$

If $\sum \vec{\tau}_{\text{ext}} = 0$ then $\vec{L}_{tot}$ = constant   (10.39)

Conservation of angular momentum can be used to find the final angular speed in terms of the initial angular speed when a zero net torque acts on a body rotating about a fixed axis, and a redistribution of mass changes the moment of inertia from I_i to I_f.

$$I_i \omega_i = I_f \omega_f = \text{constant}$$

The **parallel axis theorem** enables us to express the moment of inertia I_p through any axis parallel to the axis through the center of mass of an object.

$$I_p = I_{\text{CM}} + MD^2 \qquad (10.43)$$

M = total mass of the object
D = distance from the center-of-mass axis to the parallel axis

The **total kinetic energy of a rolling object** is the sum of the rotational kinetic energy about the center of mass and the translational kinetic energy of the center of mass.

$$K = \frac{1}{2}\left(I_{\text{CM}} + MR^2\right)\omega^2 = \frac{1}{2}I_P\omega^2 \qquad (10.44)$$

The **power supplied to a body** by a net torque is proportional to the angular speed.

$$\mathscr{P} = \tau\omega \qquad (10.31)$$

SUGGESTIONS, SKILLS, AND STRATEGIES

You should know how to calculate the moment of inertia of a system of particles about a specified axis. The technique is straightforward, and consists of applying $I = \Sigma m_i r_i^2$, where m_i is the mass of the ith particle and r_i is the distance from the axis of rotation to the particle.

    Once the moment of inertia about an axis through the center of mass I_{CM} is known, you can easily evaluate the moment of inertia about any axis parallel to the axis through the center of mass using the **parallel axis theorem**:

$$I = I_{CM} + Md^2$$

FIG. 10.1

where d is the distance between the two axes.

    For example, the moment of inertia of a solid cylinder about an axis through its center (the z axis in Figure 10.1) is given by $I_z = \frac{1}{2}MR^2$. Hence, the moment of inertia about the z' axis located a distance $d = R$ from the z axis is

$$I_{z'} = I_z + MR^2 = \tfrac{1}{2}MR^2 + MR^2 = \tfrac{3}{2}MR^2.$$

This chapter includes a discussion of the application of Newton's laws in a special situation, namely, rigid bodies in static equilibrium. It is important that you understand and follow the procedures for analyzing such problems. The following skills must be mastered in this regard:

- Recognize all external forces acting on the body, and construct an accurate free-body diagram.

- Resolve the external forces into their rectangular components, and apply the first condition of equilibrium $\sum F_x = 0$ and $\sum F_y = 0$.

- Choose a convenient origin for calculating the net torque on the body. The choice of this origin is arbitrary.

- Solve the set of simultaneous equations obtained from the two conditions of equilibrium.

The following procedure is recommended when analyzing a body in equilibrium under the action of several external forces:

- Make a sketch of the object under consideration.

- Draw a free-body diagram and label all external forces acting on the object. Try to guess the correct direction for each force. If you select an incorrect direction that leads to a negative sign in your solution for a force, do not be alarmed; this merely means that the direction of the force is the opposite of what you assumed.

- Resolve all forces into rectangular components, choosing a convenient coordinate system. Then apply the first condition for equilibrium, which balances forces. Remember to keep track of the signs of the various force components.

- Choose a convenient axis for calculating the net torque on the object. Remember that the choice of the origin for the torque equation is arbitrary; therefore, choose an origin that will simplify your calculation as much as possible. Becoming adept at this is a matter of practice.

- The first and second conditions of equilibrium give a set of linear equations with several unknowns. All that is left is to solve the simultaneous equations for the unknowns in terms of the known quantities.

The operation of the vector or cross product is used for the first time in this chapter. (Recall that the angular momentum $\vec{L}$ of a particle is defined as $\vec{L} = \vec{r} \times \vec{p}$, while torque is defined by the expression $\vec{\tau} = \vec{r} \times \vec{F}$.) You should review the cross-product operation and some of its properties.

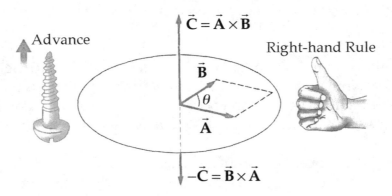

FIG. 10.2

The direction of the vector $\vec{A} \times \vec{B}$ is perpendicular to the plane formed by $\vec{A}$ and $\vec{B}$, and its sense is determined by the right-hand rule. You should practice this rule for various choices of vector pairs. You should not confuse the cross product of two vectors, which is a vector quantity, with the dot product of two vectors, which is a scalar quantity. The vector product is not commutative ($\vec{A} \times \vec{B} = -\vec{B} \times \vec{A}$) and the cross product of any vector with itself is zero.

REVIEW CHECKLIST

✓ Define the cross product (magnitude and direction) of any two vectors, $\vec{A}$ and $\vec{B}$, and state the various properties of the cross product.

✓ Apply the conservation of angular momentum principle to a body rotating about a fixed axis, in which the moment of inertia changes due to a change in the mass distribution.

✓ Describe the center of mass motion of a rigid body as it undergoes both rotation about some axis and translation in space. Note that for pure rolling motion of an object such as a sphere or cylinder, the total kinetic energy can be expressed as the sum of a rotational kinetic energy about the center of mass plus the translational energy of the center of mass.

✓ Quantitatively, the angular displacement, speed, and acceleration for a rigid body system in rotational motion are related to the distance traveled, tangential speed, and tangential acceleration. The linear quantity is calculated by multiplying the angular quantity by the radius arm for an object or point in that system.

✓ If a body rotates about a fixed axis, every particle on the body has the same angular speed and angular acceleration. For this reason, rotational motion can be simply described using these quantities. The formulas that describe angular motion are analogous to the corresponding set of formulas pertaining to linear motion.

✓ Calculate the moment of inertia I of a system of particles or a rigid body about a specific axis. Note that the value of I depends on (a) the mass distribution and (b) the axis about which the rotation occurs. The parallel-axis theorem is useful for calculating I about an axis parallel to one that goes through the center of mass.

✓ Understand the concept of torque associated with a force, noting that the torque associated with a force has a magnitude equal to the force times the moment arm. Furthermore, note that the value of the torque depends on the origin about which it is evaluated.

✓   Recognize that the work-energy theorem can be applied to a rotating rigid body. That is, the net work done on a rigid body rotating about a fixed axis equals the change in its rotational kinetic energy.

✓   Describe the two necessary conditions of equilibrium for a rigid body.

✓   Analyze problems of rigid bodies in static equilibrium using the procedures presented in Section 10.6 of the text.

ANSWERS TO SELECTED QUESTIONS

Q10.1   What is the angular speed of the second hand of a clock? What is the direction of $\vec{\omega}$ as you view a clock hanging on a vertical wall? What is the magnitude of the angular acceleration $\vec{\alpha}$ of the second hand?

Answer   The second hand of a clock turns at one revolution per minute, so

$$\omega = \frac{2\pi \, \text{rad}}{60 \, \text{s}} = 0.105 \ \text{rad/s}$$

FIG. Q10.1

The motion is clockwise, so the direction of the vector angular velocity is away from you. It turns steadily, so $\vec{\omega}$ is constant, and $\vec{\alpha}$ is zero.

Q10.5   If you see an object rotating, is there necessarily a net torque acting on it?

Answer   An object rotates with constant angular momentum when zero total torque acts on it. For example, consider the Earth; it rotates at a constant rate of once per day, but there is no net torque acting on it.

Q10.11   Why does a long pole help a tightrope walker stay balanced?

Answer   The long pole increases the tightrope walker's moment of inertia, and therefore decreases his angular acceleration, under any given torque. That gives him more time to respond, and in essence, aids his reflexes.

Q10.17 A ladder stands on the ground, leans against a wall. Would you feel safer climbing up the ladder if you were told that the ground is frictionless but the wall is rough, or that the wall is frictionless but the floor is rough? Justify your answer.

Answer The picture shows the forces on the ladder if both the wall and floor exert friction. If the floor is perfectly smooth, it can exert no frictional force to the right, to counterbalance the wall's normal force. Therefore a ladder on a smooth floor cannot stand in equilibrium. On the other hand, a smooth wall can still exert a normal force to hold the ladder in equilibrium against horizontal motion. The counterclockwise torque of this force prevents rotation about the foot of the ladder. So you should choose a rough floor.

FIG. Q10.17

SOLUTIONS TO SELECTED PROBLEMS

P10.3 An electric motor rotating a grinding wheel at 100 rev/min is switched off. The wheel then moves with a constant negative angular acceleration of magnitude 2.00 rad/s^2.

(a) During what time interval does the wheel come to rest?

(b) Through how many radians does it turn while it is slowing down?

Solution We use the rigid body under constant angular acceleration model. We are given

$$\alpha = -2.00 \text{ rad/s}^2, \ \omega_f = 0 \text{ and } \omega_i = 100 \frac{\text{rev}}{\text{min}} \left(2\pi \frac{\text{rad}}{\text{rev}}\right)\left(\frac{1 \text{ min}}{60.0 \text{ s}}\right) = 10.47 \text{ rad/s}.$$

(a) $\omega_f = \omega_i + \alpha t$: $\qquad t = \dfrac{\omega_f - \omega_i}{\alpha} = \dfrac{0 - (10.5 \text{ rad/s})}{-2.00 \text{ rad/s}^2} = 5.24 \text{ s}$ ◊

(b) $\omega_f^2 - \omega_i^2 = 2\alpha(\theta_f - \theta_i)$

$$\theta_f - \theta_i = \frac{\omega_f^2 - \omega_i^2}{2\alpha} = \frac{0 - (10.5 \text{ rad/s})^2}{2(-2.00 \text{ rad/s}^2)} = 27.4 \text{ rad} \qquad ◊$$

Note also in part (b) that since a constant acceleration is acting for time t,

$$\theta_f - \theta_i = \omega_{\text{avg}} t = \left(\frac{10.5 + 0 \text{ rad/s}}{2}\right)(5.24 \text{ s}) = 27.4 \text{ rad} \qquad ◊$$

P10.9   A disk 8.00 cm in radius rotates at a constant rate of 1 200 rev/min about its central axis.

(a)   Determine its angular speed.

(b)   Determine the tangential speed at a point 3.00 cm from its center.

(c)   Determine the radial acceleration of a point on the rim.

(d)   Determine the total distance a point on the rim moves in 2.00 s.

Solution (a) $\omega = 2\pi f = (2\pi \text{ rad/rev})\left(\dfrac{1200 \text{ rev/min}}{60 \text{ s/min}}\right) = 125.7 \text{ rad/s} = 126 \text{ rad/s}$ ◊

(b) $v = \omega R = (125.7 \text{ rad/s})(0.0300 \text{ m}) = 3.77 \text{ m/s}$ ◊

(c) $a_c = \omega^2 R = (125.7 \text{ rad/s})^2(0.0800 \text{ m}) = 1.26 \times 10^3 \text{ m/s}^2$ ◊

(d) $s = R\theta = R\omega t = (8.00 \times 10^{-2} \text{ m})(125.7 \text{ rad/s})(2.00 \text{ s}) = 20.1 \text{ m}$ ◊

P10.15   This problem describes one experimental method of determining the moment of inertia of an irregularly shaped object such as the payload for a satellite. Figure P10.15 shows a counterweight of mass m suspended by a cord wound around a spool of radius r, forming part of a turntable supporting the object. The turntable can rotate without friction. When the counterweight is released from rest, it descends through a distance h, acquiring a speed v. Show that the moment of inertia I of the rotating object (including the turntable) is

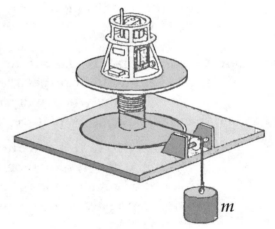

FIG. P10.15

$$mr^2\left(\dfrac{2gh}{v^2} - 1\right)$$

continued on next page

Solution　If the friction is negligible, then the energy of the counterweight-payload-turntable-Earth system is conserved as the counterweight unwinds. Each point on the cord moves at a linear speed of $v = \omega r$, where r is the radius of the spool. The energy conservation equation gives us:

$$\left(K_1 + K_2 + U_g\right)_i + W_{other} = \left(K_1 + K_2 + U_g\right)_f.$$

Solving, we have

$$0 + 0 + mgh + 0 + 0 = \frac{1}{2}mv^2 + \frac{1}{2}I\omega^2 + 0 + 0$$

$$mgh = \frac{1}{2}mv^2 + \frac{1}{2}\frac{Iv^2}{r^2}$$

$$2mgh - mv^2 = I\frac{v^2}{r^2}$$

and finally, $I = mr^2\left(\dfrac{2gh}{v^2} - 1\right)$.　◊

P10.21　Find the net torque on the wheel in Figure P10.21 about the axle through O taking $a = 10.0$ cm and $b = 25.0$ cm.

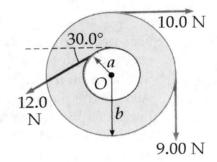

FIG. P10.21

Solution　**Conceptualize:** By examining the magnitudes of the forces and their respective lever arms, it appears that the wheel will rotate clockwise, and the net torque appears to be about 5 N·m.

Categorize: To find the net torque, we add the individual torques, remembering to apply the convention that a torque producing clockwise rotation is negative and a counterclockwise rotation is positive.

Analyze: $\sum \tau = \sum Fd = +(12.0 \text{ N})(0.100 \text{ m}) - (10.0 \text{ N})(0.250 \text{ m}) - (9.00 \text{ N})(0.250 \text{ m})$
$\sum \tau = -3.55 \text{ N·m}$　◊

(This is 3.55 N·m into the plane of the page, a clockwise torque.)

Finalize: The resulting torque has a reasonable magnitude and produces clockwise rotation as expected. Note that the 30° angle was not required for the solution since each force acted perpendicular to its lever arm. Note also that the 10-N force is to the right, but its torque is negative—that is, clockwise, just like the torque of the downward 9-N force.

P10.27 A uniform beam of mass m_b and length ℓ supports blocks with masses m_1 and m_2 at two positions as shown in Figure P10.27. The beam rests on two knife edges. For what value of x will the beam be balanced at P such that the normal force at O is zero?

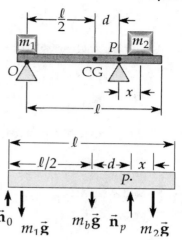

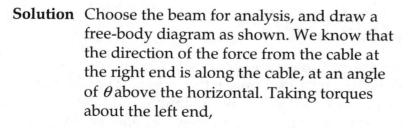

FIG. P10.27

Solution We use the rigid body in equilibrium model.

Refer to the free-body diagram, and take torques about point P.

$$\Sigma\tau_P = -n_0\left[\frac{\ell}{2}+d\right] + m_1 g\left[\frac{\ell}{2}+d\right] + m_b g d - m_2 g x = 0$$

We want to find x for which $n_0 = 0$. Let $n_0 = 0$ and solve for x.

$$m_1\left[\frac{\ell}{2}+d\right] + m_b d - m_2 x = 0 \text{ so } x = \frac{m_1}{m_2}\left[\frac{\ell}{2}+d\right] + \frac{m_b}{m_2}d$$

P10.31 A uniform sign of weight F_g and width $2L$ hangs from a light, horizontal beam hinged at the wall and supported by a cable (Fig. P10.31).

(a) Determine the tension in the cable

(b) Determine the components of the reaction force exerted by the wall on the beam in terms of F_g, d, L, and θ.

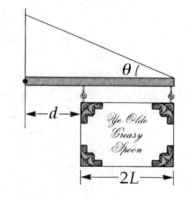

Solution Choose the beam for analysis, and draw a free-body diagram as shown. We know that the direction of the force from the cable at the right end is along the cable, at an angle of θ above the horizontal. Taking torques about the left end,

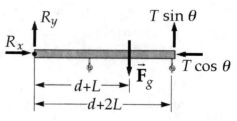

FIG. P10.31

continued on next page

$$\Sigma F_x = 0: \ +R_x - T\cos\theta = 0$$

$$\Sigma F_y = 0: \ +R_y - F_g + T\sin\theta = 0$$

$$\Sigma \tau = 0: \ R_y(0) + R_x(0) - F_g(d+L) + (0)(T\cos\theta) + (d+2L)(T\sin\theta) = 0$$

(a)   The torque equation gives $T = \dfrac{F_g(d+L)}{(d+2L)\sin\theta}.$ ◊

(b)   Now from the force equations, $R_x = \dfrac{F_g(d+L)}{(d+2L)\tan\theta}$ ◊

and $R_y = F_g - \dfrac{F_g(d+L)}{d+2L} = \dfrac{F_g L}{d+2L}.$ ◊

P10.37   Two blocks, as shown in Figure P10.37, are connected by a string of negligible mass passing over a pulley of radius 0.250 m and moment of inertia I. The block on the frictionless incline is moving up with a constant acceleration of 2.00 m/s^2.

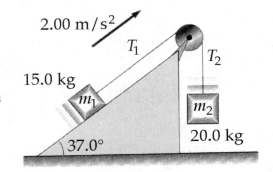

FIG. P10.37

(a)   Determine T_1 and T_2, the tensions in the two parts of the string.

(b)   Find the moment of inertia of the pulley.

Solution Conceptualize: In earlier problems, we assumed that the tension in a string was the same on either side of a pulley. Here we see that the moment of inertia changes that assumption, but we should still expect the tensions to be similar in magnitude (about the weight of each mass ~150 N), and $T_2 > T_1$ for the pulley to rotate clockwise as shown.

If we knew the mass of the pulley, we could calculate its moment of inertia, but since we only know the acceleration, it is difficult to estimate I. We at least know that I must have units of kg·m^2, and a 50-cm disk probably has a mass less than 10 kg, so I is probably less than 0.3 kg·m^2.

Categorize: For each block, we know its mass and acceleration, so we can use Newton's second law to find the net force, and from it the tension. The difference in the two tensions causes the pulley to rotate, so this net torque and the resulting angular acceleration can be used to find the pulley's moment of inertia.

continued on next page

Analyze:

(a)   Apply $\sum F = ma$ to each block to find each string tension. The forces acting on the 15-kg block are its weight, the normal support from the incline, and T_1. Taking the positive x axis as directed up the incline, $\sum F_x = ma_x$ yields: $-(m_1 g)_x + T_1 = m_1(+a)$. Substituting known values, and solving for T_1,

$$-(15.0 \text{ kg})(9.80 \text{ m/s}^2)\sin 37° + T_1 = (15.0 \text{ kg})(2.00 \text{ m/s}^2)$$
$$T_1 = 118 \text{ N} \qquad \lozenge$$

Similarly, for the counterweight, we have $\sum F_y = ma_y$ or $T_2 - m_2 g = m_2(-a)$

$$T_2 - (20.0 \text{ kg})(9.80 \text{ m/s}^2) = (20.0 \text{ kg})(-2.00 \text{ m/s}^2)$$

So, $T_2 = 156 \text{ N}$. $\qquad \lozenge$

(b)   Now for the pulley, $\sum \tau = r(T_2 - T_1) = I\alpha$. We may choose to call clockwise positive. The angular acceleration is: $\alpha = \dfrac{a}{r} = \dfrac{+2.00 \text{ m/s}^2}{0.250 \text{ m}} = +8.00 \text{ rad/s}^2$

$\sum \tau = I\alpha$: $(-118 \text{ N})(0.250 \text{ m}) + (156 \text{ N})(0.250 \text{ m}) = I(+8.00 \text{ rad/s}^2)$

$$I = \frac{9.38 \text{ N}\cdot\text{m}}{8.00 \text{ rad/s}^2} = 1.17 \text{ kg}\cdot\text{m}^2 \qquad \lozenge$$

Finalize: The tensions are close to the weight of each mass and $T_2 > T_1$ as expected. However, the moment of inertia for the pulley is about 4 times greater than expected. If the pulley is a solid disk, our result means that the pulley has a mass of 37.4 kg (about 80 lb), which means that the pulley is probably made of a dense material, like steel. This is certainly not a problem where the mass of the pulley can be ignored since the pulley has more mass than the combination of the two blocks!

P10.39 An object with a weight of 50.0 N is attached to the free end of a light string wrapped around a reel of radius 0.250 m and mass 3.00 kg. The reel is a solid disk, free to rotate in a vertical plane about the horizontal axis passing through its center. The suspended object is released 6.00 m above the floor.

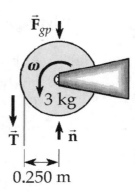

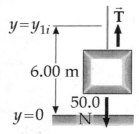

(a) Determine the tension in the string, the acceleration of the object, and the speed with which the object hits the floor.

(b) Verify your last answer by using the principle of conservation of energy to find the speed with which the object hits the floor.

FIG. P10.39

Solution **Conceptualize:** Since the rotational inertia of the reel will slow the fall of the weight, we should expect the downward acceleration to be less than g. If the reel did not rotate, the tension in the string would be equal to the weight of the object; and if the reel disappeared, the tension would be zero. Therefore, $T < mg$ for the given problem. With similar reasoning, the final speed must be less than if the weight were to fall freely:

$$v_f < \sqrt{2g\Delta y} \approx 11 \text{ m/s}.$$

Categorize: We can find the acceleration and tension using the rotational form of Newton's second law. The final speed can be found from the kinematics equation states above and from conservation of energy. Free-body diagrams will greatly assist in analyzing the forces.

Analyze:
(a) Use $\Sigma\tau = I\alpha$ to find T and a. First find I for the reel, which we assume to be a uniform disk.

$$I = \frac{1}{2}MR^2 = \frac{1}{2}3.00 \text{ kg}(0.250 \text{ m})^2 = 0.093\ 8 \text{ kg}\cdot\text{m}^2$$

The forces on it are shown, including a normal force exerted by its axle. From the diagram, we can see that the tension is the only unbalanced force causing the wheel to rotate. $\Sigma\tau = I\alpha$ becomes

continued on next page

$$n(0) + F_g(0) + T(0.250 \text{ m}) = (0.0938 \text{ kg} \cdot \text{m}^2)\left(\frac{a}{0.250 \text{ m}}\right) \qquad (1)$$

where we have applied $a_t = r\alpha$ to the point of contact between string and pulley.

The falling object has mass $m = \dfrac{F_g}{g} = \dfrac{50.0 \text{ N}}{9.80 \text{ m/s}^2} = 5.10 \text{ kg}$. ◊

For this mass, $\Sigma F_y = ma_y$ becomes

$$50.0 \text{ N} - T = (5.10 \text{ kg})a \qquad (2)$$

Note that we have defined downwards to be positive, so that positive linear acceleration of the object corresponds to positive angular acceleration of the reel. We now have our two equations in the unknowns T and a for the two connected objects. Substituting T from equation (2) into equation (1), we have

$$[50.0 \text{ N} - (5.10 \text{ kg})a](0.250 \text{ m}) = (0.0938 \text{ kg} \cdot \text{m}^2)\left(\frac{a}{0.250 \text{ m}}\right)$$

$$12.5 \text{ N} \cdot \text{m} - (1.28 \text{ kg} \cdot \text{m})a = (0.375 \text{ kg} \cdot \text{m})a$$

$12.5 \text{ N} \cdot \text{m} = a(1.65 \text{ kg} \cdot \text{m})$ or $a = 7.57 \text{ m/s}^2$ ◊

and $T = 50.0 \text{ N} - 5.10 \text{ kg}(7.57 \text{ m/s}^2) = 11.4 \text{ N}$. ◊

For the motion of the weight,
$$v_f^2 = v_i^2 + 2a(x_f - x_i) = 0^2 + 2(7.57 \text{ m/s}^2)(6.00 \text{ m})$$

$$v_f = 9.53 \text{ m/s} \quad (\text{down}) \qquad ◊$$

(b)   The work-kinetic energy theorem can take account of multiple objects more easily than Newton's second law. Like your bratty cousins, the work-kinetic energy theorem grows between visits; now it reads:

$$\left(K_1 + K_2 + U_g\right)_i = \left(K_1 + K_2 + U_g\right)_f$$

$$0 + 0 + m_1 g y_{1i} + 0 = \frac{1}{2}m_1 v_{1f}^2 + \frac{1}{2}I_2\omega_{2f}^2 + 0 + 0$$

continued on next page

Now note that $\omega = \dfrac{v}{r}$ as the string unwinds from the reel. Substituting,

$$50.0 \text{ N}(6.00 \text{ m}) = \frac{1}{2}(5.10 \text{ kg})v_f^2 + \frac{1}{2}\left(0.093\ 8 \text{ kg} \cdot \text{m}^2\right)\left(\frac{v_f}{0.250 \text{ m}}\right)^2$$

$$300 \text{ N} \cdot \text{m} = \frac{1}{2}(5.10 \text{ kg})v_f^2 + \frac{1}{2}(1.50 \text{ kg})v_f^2$$

$$v_f = \sqrt{\frac{2(300 \text{ N} \cdot \text{m})}{6.60 \text{ kg}}} = 9.53 \text{ m/s} \qquad \lozenge$$

Finalize: As we should expect, both methods give the same final speed for the falling object, but the energy method is simpler. The acceleration is less than g, and the tension is less than the object's weight as we predicted. Now that we understand the effect of the reel's moment of inertia, this problem solution could be applied to solve other real-world pulley systems with masses that should not be ignored.

P10.41   The position vector of a particle of mass 2.00 kg is given as a function of time by $\vec{\mathbf{r}} = \left(6.00\hat{\mathbf{i}} + 5.00t\hat{\mathbf{j}}\right) \text{ m}$. Determine the angular momentum of the particle about the origin, as a function of time.

Solution   The velocity of the particle is $\vec{\mathbf{v}} = \dfrac{d\vec{\mathbf{r}}}{dt} = \dfrac{d}{dt}\left(6.00\hat{\mathbf{i}} \text{ m} + 5.00t\hat{\mathbf{j}} \text{ m}\right) = 5.00\hat{\mathbf{j}} \text{ m/s}$. The angular momentum is

$$\vec{\mathbf{L}} = \vec{\mathbf{r}} \times \vec{\mathbf{p}} = m\vec{\mathbf{r}} \times \vec{\mathbf{v}} = (2.00 \text{ kg})\left(6.00\hat{\mathbf{i}} \text{ m} + 5.00t\hat{\mathbf{j}} \text{ m}\right) \times 5.00\hat{\mathbf{j}} \text{ m/s}$$

$$\vec{\mathbf{L}} = \left(60.0 \text{ kg} \cdot \text{m}^2/\text{s}\right)\hat{\mathbf{i}} \times \hat{\mathbf{j}} + \left(50.0t \text{ kg} \cdot \text{m}^2/\text{s}\right)\hat{\mathbf{j}} \times \hat{\mathbf{j}}$$

$$\vec{\mathbf{L}} = \left(60.0\hat{\mathbf{k}}\right) \text{ kg} \cdot \text{m}^2/\text{s}, \text{ constant in time} \qquad \lozenge$$

P10.47   A 60.0-kg woman stands at the rim of a horizontal turntable having a moment of inertia of $500 \text{ kg} \cdot \text{m}^2$ and a radius of 2.00 m. The turntable is initially at rest and is free to rotate about a frictionless, vertical axle through its center. The woman then starts walking around the rim clockwise (as viewed from above the system) at a constant speed of 1.50 m/s relative to the Earth.

continued on next page

(a)   In what direction and with what angular speed does the turntable rotate?

(b)   How much work does the woman do to set herself and the turntable into motion?

Solution The table rotates in a direction opposite to that in which the woman walks. We use the angular momentum version of the isolated system model. There are no external torques acting on the woman-turntable system; therefore, from conservation of angular momentum, we have $L_f = L_i = 0$.

(a)   Therefore, $L_f = I_w \omega_w + I_t \omega_t = 0$ and $\omega_t = -\dfrac{I_w}{I_t} \omega_w$. Solving,

$$\omega_t = -\left(\frac{m_w r^2}{I_t}\right)\left(\frac{v_w}{r}\right) = -\frac{(60.0 \text{ kg})(2.00 \text{ m})(1.50 \text{ m/s})}{500 \text{ kg} \cdot \text{m}^2}$$

$$\omega_t = -0.360 \text{ rad/s} = 0.360 \text{ rad/s CCW} \qquad \lozenge$$

(b)   Work done $= \Delta K$: $W = K_f - 0 = \dfrac{1}{2} m_{woman} v_{woman}^2 + \dfrac{1}{2} I_{table} \omega_{table}^2$

$$W = \frac{1}{2}(60.0 \text{ kg})(1.50 \text{ m/s})^2 + \frac{1}{2}(500 \text{ kg} \cdot \text{m}^2)(0.360 \text{ rad/s})^2 = 99.9 \text{ J} \qquad \lozenge$$

Related Questions:

(a)   Why is the angular momentum of the woman-turntable system conserved?

(b)   Why is the mechanical energy of this system not conserved?

(c)   Is the linear momentum of this system conserved?

Answers:

(a)   Because the axle exerts no torque on the woman-plus-turntable system; only torques from outside the system can change the total angular momentum.

continued on next page

(b)   The internal forces, of the woman pushing backward on the turntable and of the turntable pushing forward on the woman, both do positive work, converting chemical into kinetic energy.

(c)   No. If the woman starts walking north, she pushes south on the turntable. Its axle holds it still against linear motion by pushing north on it, and this outside force delivers northward linear momentum into the system.

P10.53   A cylinder of mass 10.0 kg rolls without slipping on a horizontal surface. At a certain instant its center of mass has a speed of 10.0 m/s.

(a)   Determine the translational kinetic energy of its center of mass.

(b)   Determine the rotational kinetic energy about its center of mass.

(c)   Determine its total energy.

Solution (a) $K_{trans} = \dfrac{1}{2}mv_{CM}{}^2 = \dfrac{1}{2}(10.0 \text{ kg})(10.0 \text{ m/s})^2 = 500 \text{ J}$ ◊

(b)   Call the radius of the cylinder R. An observer at the center sees the rough surface and the circumference of the cylinder moving at 10.0 m/s, so the angular speed of the cylinder is:

$$\omega = \frac{v_{CM}}{R} = \frac{10.0 \text{ m/s}}{R}$$

The moment of inertia about an axis through the center of mass is:

$$I_{CM} = \frac{1}{2}mR^2 \text{ so}$$

$$K_{rot} = \frac{1}{2}I_{CM}\omega^2 = \left(\frac{1}{2}\right)\left[\frac{1}{2}(10.0 \text{ kg})R^2\right]\left(\frac{10.0 \text{ m/s}}{R}\right)^2 = 250 \text{ J}$$ ◊

(c)   We can now add up the total energy: $K_{tot} = 500 \text{ J} + 250 \text{ J} = 750 \text{ J}$ ◊

P10.59 A long uniform rod of length L and mass M is pivoted about a horizontal, frictionless pin passing through one end. The rod is released from rest in a vertical position as shown in Figure P10.59. At the instant the rod is horizontal,

(a) find its angular speed.

(b) find the magnitude of its angular acceleration.

(c) find the x and y components of the acceleration of its center of mass.

(d) find the components of the reaction force at the pivot.

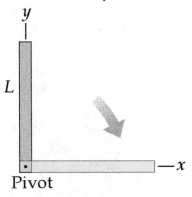

FIG. P10.59

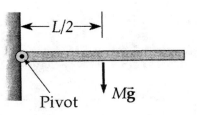

FIG. P10.59(a)

Solution (a) Since only conservative forces are acting on the bar, use conservation of energy of the bar-Earth system: $\Delta K + \Delta U = 0$ or $K_f - K_i + U_f - U_i = 0$. For evaluation of its gravitational energy, a rigid body can be modeled as a particle at its center of mass. Take the zero configuration for potential energy with the bar horizontal. Under these conditions $U_f = 0$ and $U_i = \dfrac{MgL}{2}$. Using the equation above,

$$\left(\frac{1}{2}I\omega_f{}^2 - 0\right) + \left(0 - \frac{1}{2}MgL\right) = 0$$

and $\omega_f = \sqrt{\dfrac{MgL}{I}}$. For a bar rotating about an axis through one end, $I = \dfrac{ML^2}{3}$.

Therefore, $\omega_f = \sqrt{\dfrac{MgL}{(1/3)ML^2}} = \sqrt{\dfrac{3g}{L}}$. ◊

Note that we have chosen clockwise rotation as positive.

continued on next page

(b) $\sum \tau = I\alpha$: $Mg\left(\dfrac{L}{2}\right) = \left(\dfrac{1}{3}ML^2\right)\alpha$ and $\alpha = \dfrac{3g}{2L}$ ◊

(c) $a_x = -a_c = -r\omega_f^{\,2} = -\left(\dfrac{L}{2}\right)\left(\dfrac{3g}{L}\right) = -\dfrac{3g}{2}$ ◊

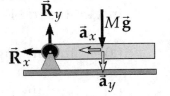

Since this is **centripetal** acceleration, it is
directed along the **negative** horizontal.

FIG. P10.59(c)

$$a_y = -a_t = -r\alpha = -\dfrac{L}{2}\alpha = -\dfrac{3g}{4}$$ ◊

(d)   Using $\sum \vec{F} = m\vec{a}$, we have $R_x = Ma_x = -\dfrac{3Mg}{2}$ in the **negative** direction   ◊

$$R_y - Mg = Ma_y \text{ so } R_y = M(g + a_y) = M\left(g - \dfrac{3g}{4}\right) = \dfrac{Mg}{4}.$$ ◊

One could say that the motion of the rod is surprisingly rapid. Its angular
acceleration is so large that the tangential acceleration of its moving end is
$r\alpha = L\left(\dfrac{3g}{2L}\right) = \dfrac{3g}{2}$ larger than the acceleration of gravity. Its angular speed
is so large that the pin must exert a considerable centripetal force on the
rod, three times larger than the vertical force component exerted by the
pin.

P10.73   A force acts on a rectangular cabinet weighing 400 N, as in Figure P10.73.

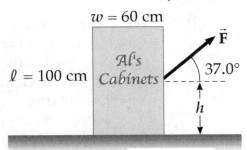

FIG. P10.73

(a)   Assuming that the cabinet slides with constant speed when $F = 200$ N and $h = 0.400$ m, find the coefficient of kinetic friction and the position of the resultant normal force.

(b)   Taking $F = 300$ N, find the value of h for which the cabinet just begins to tip.

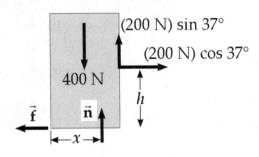

FIG. P10.73(a)

Solution   (a)   Think of the normal force as acting at a distance x from the lower left corner. Moving with constant speed, the cabinet is in equilibrium:

$$\sum F_x = 0: \quad -f + (200 \text{ N})\cos(37.0°) = 0$$

$$\sum F_y = 0: \quad -400 \text{ N} + n + (200 \text{ N})\sin(37.0°) = 0$$

These equations tell us that $f = 160$ N and $n = 280$ N,

so $\mu_k = \dfrac{f}{n} = 0.571$.   ◊

Take torques about the lower left corner; $\Sigma\tau = 0$ gives

$$-(400 \text{ N})(30 \text{ cm}) + nx + (200 \text{ N})(60 \text{ cm})\sin(37°) - (200 \text{ N})(40 \text{ cm})\cos(37°) = 0$$

Substituting $n = 280$ N (and converting cm to m, and back) gives

$$x = \frac{120 \text{ N}\cdot\text{m} - 72.2 \text{ N}\cdot\text{m} + 63.9 \text{ N}\cdot\text{m}}{280 \text{ N}} = 39.9 \text{ cm}$$ ◊

continued on next page

(b)   When the cabinet is just about to tip, the normal force is located at the lower right corner, and $\Sigma\tau = 0$ is still true.

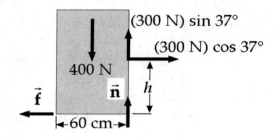

FIG. P10.73(b)

Because most of the forces are directed through the lower right corner, we choose to take torques about that point. This leaves only two forces to deal with.

$$\Sigma\tau = 0: \quad -(300 \text{ N})(h)\cos(37.0°) + (400 \text{ N})(30.0 \text{ cm}) = 0$$

Solving for h, $h = \dfrac{120 \text{ N} \cdot \text{m}}{240 \text{ N}} = 50.1 \text{ cm}.$ ◊

Chapter 11

Gravity, Planetary Orbits, and the Hydrogen Atom

NOTES FROM SELECTED CHAPTER SECTIONS

Section 11.1   Newton's Law of Universal Gravitational Revisited

Newton's law of gravitation states that every particle in the universe attracts every other particle with a force that is directly proportional to the product of their masses and inversely proportional to the square of their distance of separation. The gravitational force always exists between two particles regardless of the medium that separates them.

The gravitational force exerted by a finite-size, spherically symmetric mass distribution on a particle outside the sphere is the same as if the entire mass of the sphere were concentrated at its center.

Section 11.3   Kepler's Laws

Kepler deduced the following three empirical laws as they apply to our solar system.

- All planets move in elliptical orbits with the Sun at one of the focal points.

- The radius vector from the Sun to any planet sweeps out equal areas in equal times.

- The square of the orbital period of any planet is proportional to the cube of the semimajor axis for the elliptical orbit.

Kepler's second law is a consequence of the central nature of the gravitational force, which leads to conservation of angular momentum. Kepler's third law follows from the inverse square nature of the gravitational force.

Section 11.4 Energy Considerations in Planetary and Satellite Motion

Both the total energy and the total angular momentum of a planet-Sun system are **constants of the motion**.

The **gravitational potential energy** for any pair of particles varies as $1/r$. The potential energy is negative since the force is one of attraction and the potential energy is taken to be zero when the distance of separation between the two particles is infinite. The absolute value of the potential energy is the **binding** energy of the system.

In the case of a bound system (closed orbits), the total energy must be negative. The kinetic energy in the case of a circular orbit, is equal to one-half the magnitude of the potential energy.

The **escape speed** for an object projected from the Earth is independent of the mass of the object and it is independent of the direction of the initial speed (provided that the trajectory does not intersect Earth.) Also, when the initial speed is equal to the escape speed, the total energy is equal to zero.

Section 11.5 Atomic Spectra and the Bohr Theory of Hydrogen

It is important to understand the behavior of the hydrogen atom as an atomic system for the following reasons:

* The quantum numbers used to characterize the allowed states of hydrogen can be used to describe the allowed states of more complex atoms. This enables us to understand the periodic table of the elements, which is one of the greatest triumphs of quantum mechanics.

* The hydrogen atom is an ideal system for performing precise tests of theory against experiment and for improving our overall understanding of atomic structure.

* Much of what is learned about the hydrogen atom with its single electron can be extended to such single-electron ions as He^+ and Li^{2+}, which are hydrogen-like in their atomic structure.

* The basic ideas about atomic structure must be well understood before we attempt to deal with the complexities of molecular structures and the electronic structure of solids.

The basic assumptions of the Bohr theory as it applies to the hydrogen atom are as follows:

- The electron moves in circular orbits about the proton under the influence of the Coulomb force of attraction. In this case, the Coulomb force is the centripetal force.

- Only certain electron orbits are stable. These are orbits in which the hydrogen atom does not emit energy in the form of radiation. Hence, the total energy of the atom remains constant.

- Radiation is emitted by the hydrogen atom when the electron "jumps" from a more energetic initial state to a lower state. The "jump" cannot be visualized or treated classically. The frequency, f, of the radiation emitted in the jump is related to the change in the atom's energy. The frequency of the emitted radiation is

$$E_i - E_f = hf$$

where E_i is the energy of the initial state, E_f is the energy of the final state, h is Planck's constant, and $E_i > E_f$.

- The size of the allowed electron orbits is determined by a condition imposed on the electron's orbital angular momentum: The allowed orbits are those for which the electron's orbital angular momentum about the nucleus is an integral multiple of $\hbar = \dfrac{h}{2\pi}$.

$$mvr = n\hbar \text{ where } n = 1, 2, 3, \ldots .$$

The lowest stationary state of an electron is called the ground state. The minimum energy required to ionize an atom (remove an electron in the ground state from the influence of the proton) is called the ionization energy.

According to the correspondence principle, quantum mechanics is in agreement with classical mechanics when the quantum numbers are very large.

EQUATIONS AND CONCEPTS

The **universal law of gravity** states that any pair of particles attract each other with a force that is proportional to the product of their masses and inversely proportional to the square of their separation.

$$F_g = G \frac{m_1 m_2}{r^2} \tag{11.1}$$

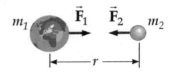

The constant G is called the **universal constant of gravity**.

$$G = 6.673 \times 10^{-11} \frac{\text{N} \cdot \text{m}^2}{\text{kg}^2} \tag{11.2}$$

The **gravitational force** $\vec{F}$ can be expressed in vector form. The unit vector $\hat{r}_{12}$ is directed from m_1 to m_2, and $\vec{F}$ is the force on m_2 due to m_1. According to Newton's third law, $\vec{F}_{12} = -\vec{F}_{21}$. *The gravitational force exerted by a finite-sized spherically symmetric mass distribution on a particle outside the distribution is the same as if the entire mass of the distribution were concentrated at its center.*

$$\vec{F}_{12} = -\frac{G m_1 m_2}{r^2} \hat{r}_{12} \tag{11.3}$$

The **gravitational field at a distance r from the center of the Earth** points radially inward toward the center of the Earth ($\hat{r}$ is a unit vector pointing radially outward from the Earth). *Over a small region near the Earth's surface, $\vec{g}$ is an approximately uniform downward field.*

$$\vec{g} = \frac{\vec{F}_g}{m} = -\frac{G M_E}{r^2} \hat{r} \tag{11.5}$$

Kepler's first law states that all planets move in elliptical orbits with the Sun at one focus (see figure at right). *This law is a direct result of the inverse square nature of the gravitational force. The semimajor axis has length a and the semiminor axis has length b.*

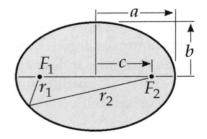

Kepler's second law states that the radius vector from the Sun to any planet sweeps out equal areas in equal times. *This result applies in the case of any central force and implies conservation of angular momentum.*

$$\frac{dA}{dt} = \frac{L}{2M_p} = \text{constant} \qquad (11.6)$$

Kepler's third law states that the square of the orbital period of any planet is proportional to the cube of the semimajor axis of the elliptical orbit. *Kepler's third law is valid for circular and elliptical orbits and the constant of proportionality is independent of the mass of the planet orbiting the Sun.* In the case of a satellite of a planet (e.g. the moon orbiting the earth), M_S is replaced by the mass of the planet and the constant becomes K_E, which has a different value.

$$T^2 = \left(\frac{4\pi^2}{GM_S}\right)a^3 = K_S a^3 \qquad (11.7)$$

$$K_S = 2.97 \times 10^{-19} \ \text{s}^2/\text{m}^3$$

The **total energy, *E*, of a two-body system**, when the bodies are separated by a distance, *r*, is the sum of the kinetic energy of the orbiting mass *m*, and the potential energy of the system, where mass *M* is assumed to be at rest in an inertial frame.

$$E = \frac{1}{2}mv^2 - G\frac{Mm}{r} \qquad (11.8)$$

When the kinetic and potential energy contributions are combined, one finds that the total energy *E* is negative as given by Eqs. 13.10 and 13.11. This arises from the fact that the (positive) kinetic energy is equal to half the magnitude of the (negative) potential energy in the case of a *circular* orbit ($a = r = \text{constant}$).

$$E = -\frac{GMm}{2r} \ \text{(circular orbits)} \qquad (11.10)$$

$$E = -\frac{GMm}{2a}v \ \text{(elliptical orbits)} \qquad (11.11)$$

In Equation 13.10 and 13.11, $M \gg m$, and assumed to be at rest.

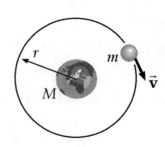

The **escape speed** is defined as the **minimum** speed a body must have, when projected from the Earth whose mass is M_E and radius is R_E, in order to escape the Earth's gravitational field (that is, to just reach $r = \infty$ with zero speed). Note that v_{esc} does not depend on the mass of the projected body. This equation can be applied to any object projected from any planet, by substituting M_E with M, and substituting R_E with R.

$$v_{esc} = \sqrt{\frac{2GM_E}{R_E}} \tag{11.13}$$

The **spectral lines of the Balmer series** which appear in the visible region of the hydrogen emission spectrum have wavelengths which can be calculated from an empirical equation.

$$\frac{1}{\lambda} = R_H\left(\frac{1}{2^2} - \frac{1}{n^2}\right) \tag{11.16}$$

where $n = 3, 4, 5, \ldots$

$R_H = 1.097\ 373\ 2 \times 10^7\ \text{m}^{-1}$

The **total energy of the hydrogen atom** is the sum of the potential energy $\left(U_e = \frac{-k_e e^2}{r}\right)$ and the kinetic energy. *Note that the total energy is negative, which is indicative of a bound electron-proton system.*

$$E = -\frac{k_e e^2}{2r} \tag{11.21}$$

Radii of the allowed orbits can be obtained by combining the equation for quantization of the orbital angular momentum of the electron with the equation for kinetic energy.

$$r_n = \frac{n^2 \hbar^2}{m_e k_e e^2} \tag{11.22}$$

where $n = 1, 2, 3, \ldots$

Quantization of energy values for the electron are a result of quantization of the values of the electron orbit radii (as assumed in the Bohr theory).

$$E_n = -\frac{13.606}{n^2}\ \text{eV} \tag{11.25}$$

where $n = 1, 2, 3, \ldots$

The **frequency and wavelength** of the emitted radiation can be calculated in terms of the initial and final quantum numbers. *Radiation is emitted by an atom when the electron makes a transition from an initial state to a final state of different orbital radius.*

$$\frac{1}{\lambda} = R_H\left(\frac{1}{n_f^2} - \frac{1}{n_i^2}\right) \tag{11.28}$$

and $f = \frac{c}{\lambda}$

REVIEW CHECKLIST

✓   State Kepler's three laws of planetary motion and recognize that the laws are empirical in nature; that is, they are based upon astronomical data.

✓   Describe the nature of Newton's universal law of gravity, and the method of deriving Kepler's third law $\left(T^2 \propto r^3\right)$ from this law for circular orbits. Recognize that Kepler's second law is a consequence of conservation of angular momentum and the central nature of the gravitational force.

✓   Describe the total energy of a planet or Earth satellite moving in a circular orbit about a large body located at the center of motion. Note that the total energy is negative, as it must be for any closed orbit.

✓   Understand the meaning of escape speed, and know how to obtain the expression for v_{esc} using the principle of conservation of energy.

✓   Describe the Bohr model of the hydrogen atom. Relate the basic assumptions and the observed spectral lines.

✓   Calculate the frequency and wavelength of the radiation that is emitted when an electron in hydrogen undergoes a transition between energy levels with different quantum numbers.

ANSWERS TO SELECTED QUESTIONS

Q11.5   Explain why it takes more fuel for a spacecraft to travel from the Earth to the Moon than for the return trip. Estimate the difference.

Answer   The mass and radius of the Earth and Moon, and the distance between the two are

$$M_E = 5.98 \times 10^{24} \text{ kg}, \ R_E = 6.37 \times 10^6 \text{ m},$$

$$M_M = 7.36 \times 10^{22} \text{ kg}, \ R_M = 1.75 \times 10^6 \text{ m, and } d = 3.84 \times 10^8 \text{ m}.$$

To travel between the Earth and the Moon, a distance d, a rocket engine must boost the spacecraft over the point of zero total gravitational field in between. Call x the distance of this point from Earth. To cancel, the Earth and Moon must here produce equal fields:

continued on next page

$$\frac{GM_E}{x^2} = \frac{GM_M}{(d-x)^2}.$$

Isolating x, $\frac{M_E}{M_M}(d-x)^2 = x^2$ and $\sqrt{\frac{M_E}{M_M}}(d-x) = x$.

Thus, $x = \frac{d\sqrt{M_E}}{\sqrt{M_M} + \sqrt{M_E}} = 3.46 \times 10^8$ m

and $(d-x) = \frac{d\sqrt{M_M}}{\sqrt{M_M} + \sqrt{M_E}} = 3.83 \times 10^7$ m.

In general, we can ignore the gravitational pull of the far planet when we're close to the other; our results will still be approximately correct. Thus the approximate energy difference between point x and the Earth's surface, is:

$$\frac{\Delta E_{E \to x}}{m} = \frac{-GM_E}{x} - \frac{-GM_E}{R_E}.$$

Similarly, flying from the moon: $\frac{\Delta E_{M \to x}}{m} = \frac{-GM_M}{d-x} - \frac{-GM_M}{R_M}.$

Taking the ratio of the energy terms, $\dfrac{\Delta E_{E \to x}}{\Delta E_{M \to x}} = \dfrac{\frac{M_E}{x} - \frac{M_E}{R_E}}{\frac{M_M}{d-x} - \frac{M_M}{R_M}} \approx 23.0.$

This would also be the minimum fuel ratio, if the spacecraft was driven by a method other than rocket exhaust. If rockets were used, the total fuel ratio would be much larger.

Q11.7   Why don't we put a geosynchronous weather satellite in orbit around the 45th parallel? Wouldn't this be more useful in the United States than one in orbit around the equator?

Answer   While a satellite in orbit above the 45th parallel might be more useful, it isn't possible. The center of a satellite orbit must be the center of the Earth, since that is the force center for the gravitational force. If the satellite is north of the plane of the equator for part of its orbit, it must be south of the equatorial plane for the rest.

Q11.9 At what position in its elliptical orbit is the speed of a planet a maximum? At what position is the speed a minimum?

Answer At the planet's closest approach to the Sun—its perihelion—the system's potential energy is at its minimum (most negative), so that the planet's kinetic energy and speed take their maximum values. At the planet's greatest separation from the Sun—its aphelion—the gravitational energy is maximum, and the speed is minimum.

Q11.11 In his 1798 experiment, Cavendish was said to have "weighed the Earth." Explain this statement.

Answer The Earth creates a gravitational field at its surface according to $g = \dfrac{GM_E}{R_E^2}$. The factors g and R_E were known, so as soon as Cavendish measured G, he could compute the mass of the Earth.

SOLUTIONS TO SELECTED PROBLEMS

P11.3 In introductory physics laboratories, a typical Cavendish balance for measuring the gravitational constant G uses lead spheres of masses 1.50 kg and 15.0 g whose centers are separated by about 4.50 cm. Calculate the gravitational force between these spheres, treating each as a particle located at the center of the sphere.

Solution $F = \dfrac{Gm_1m_2}{r^2} = \dfrac{\left(6.67 \times 10^{-11}\ \text{N} \cdot \text{m}^2/\text{kg}^2\right)(1.50\ \text{kg})(0.0150\ \text{kg})}{\left(4.50 \times 10^{-2}\ \text{m}\right)^2}$

$F = 7.41 \times 10^{-10}\ \text{N} = 741\ \text{pN}$ toward the other sphere. ◊

This is the force that each sphere has exerted on it by the other. In the space literally "between the spheres," no force acts because no object is there to feel a force.

P11.7   The free-fall acceleration on the surface of the Moon is about one-sixth that on the surface of the Earth. Assuming that the radius of the Moon is about $0.250R_E$, find the ratio of their average densities, $\dfrac{\rho_{\text{Moon}}}{\rho_{\text{Earth}}}$.

Solution   The gravitational field at the surface of the Earth or Moon is given by

$$g = \frac{GM}{R^2}.$$

The expression for density is $\rho = \dfrac{M}{V} = \dfrac{M}{\frac{4}{3}\pi R^3}$

so $$M = \frac{4}{3}\pi\rho R^3$$

and $$g = \frac{G\frac{4}{3}\pi\rho R^3}{R^2} = G\frac{4}{3}\pi\rho R.$$

Noting that this equation applies to both the Moon and the Earth, and dividing the two equations,

$$\frac{g_M}{g_E} = \frac{G\frac{4}{3}\pi\rho_M R_M}{G\frac{4}{3}\pi\rho_E R_E} = \frac{\rho_M R_M}{\rho_E R_E}$$

Substituting, $\dfrac{1}{6} = \dfrac{\rho_M}{\rho_E}\left(\dfrac{1}{4}\right)$ and $\dfrac{\rho_M}{\rho_E} = \dfrac{4}{6} = \dfrac{2}{3}$.   ◊

P11.11   Compute the magnitude and direction of the gravitational field at a point P on the perpendicular bisector of the line joining two objects of equal masses separated by a distance $2a$ as shown in Figure P11.11.

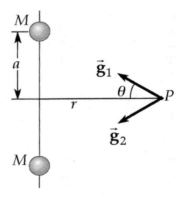

Solution   We must add the vector fields created by each mass. In equation form, $\vec{g} = \vec{g}_1 + \vec{g}_2$ where

$$\vec{g}_1 = \frac{GM}{r^2 + a^2} \text{ to the left and upward at } \theta$$

and $$\vec{g}_2 = \frac{GM}{r^2 + a^2} \text{ to the left and downward at } \theta.$$

FIG. P11.11

continued on next page

Therefore, $\vec{g} = \dfrac{GM}{r^2+a^2}\cos\theta(-\hat{i}) + \dfrac{GM}{r^2+a^2}\sin\theta(\hat{j}) + \dfrac{GM}{r^2+a^2}\cos\theta(-\hat{i}) + \dfrac{GM}{r^2+a^2}\sin\theta(-\hat{j})$

$$\vec{g} = \frac{2GM}{r^2+a^2}\frac{r}{\sqrt{r^2+a^2}}(-\hat{i}) + 0\hat{j} = \frac{-2GMr}{(r^2+a^2)^{3/2}}\hat{i}$$ ◊

The net field points toward the midpoint between the masses. It is directly proportional to M. It approaches zero both as $r \to 0$ and as $r \to \infty$.

P11.15   Io, a satellite of Jupiter, has an orbital period of 1.77 days and an orbital radius of 4.22×10^5 km. From these data, determine the mass of Jupiter.

Solution   We use the particle under a net force and the particle in uniform circular motion models. The gravitational force of Jupiter on Io provides the centripetal acceleration of Io.

$$\sum F_{Io} = M_{Io}a : \frac{GM_J M_{Io}}{r^2} = \frac{M_{Io}v^2}{r} = \frac{M_{Io}}{r}\left(\frac{2\pi r}{T}\right)^2 = \frac{4\pi^2 r M_{Io}}{T^2}$$

Thus, $M_J = \dfrac{4\pi^2 r^3}{GT^2} = \dfrac{4\pi^2(4.22\times10^8 \text{ m})^3}{(6.67\times10^{-11} \text{ N}\cdot\text{m}^2/\text{kg}^2)(1.77 \text{ d})^2}\left(\dfrac{1 \text{ d}}{86\,400 \text{ s}}\right)^2\left(\dfrac{\text{N}\cdot\text{s}^2}{\text{kg}\cdot\text{m}}\right)$

and $M_J = 1.90 \times 10^{27}$ kg .   ◊

P11.17   Plaskett's binary system consists of two stars that revolve in a circular orbit about a center of mass midway between them. Therefore, the masses of the two stars are equal (Figure P11.17). Assuming that the orbital velocity of each star is 220 km/s and the orbital period of each is 14.4 days. Find the mass M of each star. (For comparison, the mass of our Sun is 1.99×10^{30} kg.)

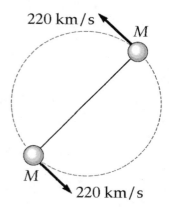

FIG. P11.17

Solution   **Conceptualize:** From the given data, it is difficult to estimate a reasonable answer to this problem without working through the details to actually solve it. A reasonable guess might be that each star has a mass larger than our Sun because fourteen days is short compared to the orbital periods of all the Sun's planets.

continued on next page

Categorize: The only force acting on the two stars is the central gravitational force of attraction that results in a centripetal acceleration. When we solve Newton's 2nd law, we can find the unknown mass in terms of the variables given in the problem.

Analyze: Applying Newton's 2nd Law, $\sum F = ma$ yields $F_g = ma_c$ for each star:

$$\frac{GMM}{(2r)^2} = \frac{Mv^2}{r} \text{ so } M = \frac{4v^2 r}{G}.$$

We can write r in terms of the period, T, by considering the time and distance of one complete cycle. The distance traveled in one orbit is the circumference of the stars' common orbit , so $2\pi r = vT$. Therefore

$$M = \frac{4v^2 r}{G} = \left(\frac{4v^2}{G}\right)\left(\frac{vT}{2\pi}\right) = \frac{2v^3 T}{\pi G}$$

$$M = \frac{2(220 \times 10^3 \text{ m/s})^3 (14.4 \text{ d})(86\ 400 \text{ s/d})}{\pi\left(6.67 \times 10^{-11} \text{ N} \cdot \text{m}^2/\text{kg}^2\right)} = 1.26 \times 10^{32} \text{ kg} \qquad \lozenge$$

Finalize: The mass of each star is about 63 solar masses, much more than our initial guess! A quick check in an astronomy book reveals that stars over 8 solar masses are considered to be **heavyweight** stars, and astronomers estimate that the maximum theoretical limit is about 100 solar masses before a protostar fissions as it forms. So these two stars are exceptionally massive!

P11.21   After our Sun exhausts its nuclear fuel, its ultimate fate may be to collapse to a **white dwarf** state, in which it has approximately the same mass it has now but a radius equal to the radius of the Earth. Calculate

(a)   the average density of the white dwarf,

(b)   the free-fall acceleration, and

(c)   the gravitational potential energy associated with a 1.00-kg object at its surface.

continued on next page

Solution (a) $\rho = \dfrac{M_s}{V} = \dfrac{M_s}{\left(\frac{4}{3}\right)\pi R_E^3} = \dfrac{1.99 \times 10^{30} \text{ kg}}{\left(\frac{4}{3}\right)\pi (6.37 \times 10^6 \text{ m})^3} = 1.84 \times 10^9 \text{ kg/m}^3$ ◊

(This is on the order of 1 million times the density of concrete!)

(b)   For an object of mass m on its surface, $mg = \dfrac{GM_s m}{R_E^2}$. Thus,

$g = \dfrac{GM_s}{R_E^2} = \dfrac{(6.67 \times 10^{-11} \text{ N} \cdot \text{m}^2/\text{kg}^2)(1.99 \times 10^{30} \text{ kg})}{(6.37 \times 10^6 \text{ m})^2} = 3.27 \times 10^6 \text{ m/s}^2$ ◊

(This acceleration is on the order of 1 million times more than g_{Earth})

(c)   Relative to $U_g = 0$ at infinity,

$U_g = \dfrac{-GM_s m}{R_E} = \dfrac{(-6.67 \times 10^{-11} \text{ N} \cdot \text{m}^2/\text{kg}^2)(1.99 \times 10^{30} \text{ kg})(1 \text{ kg})}{(6.37 \times 10^6 \text{ m})}$

$U_g = -2.08 \times 10^{13} \text{ J}$ ◊

(Such a large potential energy could yield a big gain in kinetic energy with even small changes in height. For example, dropping the 1.00-kg object from 1.00 m would result in a final velocity of 2 560 m/s.)

P11.25   A space probe is fired as a projectile upward from the Earth's surface with an initial speed of 2.00×10^4 m/s. What will its speed be when it is very far from the Earth? Ignore friction and the rotation of the Earth.

Solution   We apply the work-kinetic energy theorem between an initial point just after the payload is fired off at the Earth's surface, and a final point when it is coasting along far away.

$$K_i + U_i + W_{\text{other forces}} - f_k d = K_f + U_f$$

$$\frac{1}{2} m_{\text{probe}} v_i^2 - \frac{GM_{\text{Earth}} m_{\text{probe}}}{r_i} = \frac{1}{2} m_{\text{probe}} v_f^2 - \frac{GM_E m_{\text{probe}}}{r_f}$$

The reciprocal of the final distance is negligible compared with the reciprocal of the original distance from the Earth's center.

continued on next page

$$\frac{1}{2}v_i{}^2 - \frac{GM_{\text{Earth}}}{R_{\text{Earth}}} = \frac{1}{2}v_f{}^2 - 0$$

We solve for the final speed, and substitute values tabulated on the endpapers:

$$v_f{}^2 = v_i{}^2 - \frac{2GM_{\text{Earth}}}{R_{\text{Earth}}}$$

$$= \left(2.00 \times 10^4 \text{ m/s}\right)^2 - \frac{2\left(6.67 \times 10^{-11} \text{ N}\cdot\text{m}^2/\text{kg}^2\right)\left(5.98 \times 10^{24} \text{ kg}\right)}{6.37 \times 10^6 \text{ m}}$$

$$v_f^2 = 2.75 \times 10^8 \text{ m}^2/\text{s}^2 \qquad \text{and} \qquad v_f = 1.66 \times 10^4 \text{ m/s}. \qquad \Diamond$$

P11.27   A "treetop satellite" (Fig. P11.27) moves in a circular orbit just above the surface of a planet, assumed to offer no air resistance. Show that its orbital speed v and the escape speed from the planet are related by the expression $v_{\text{esc}} = \sqrt{2}v$.

Solution   Call M the mass of the planet and R its radius. For the orbiting "treetop satellite,"

$$\sum F = ma \text{ becomes } \frac{GMm}{R^2} = \frac{mv^2}{R} \text{ or } v = \sqrt{\frac{GM}{R}}.$$

FIG. P11.27

If the object is launched with escape velocity, applying conservation of energy to the object-Earth system gives

$$\frac{1}{2}mv_{\text{esc}}^2 - \frac{GMm}{R} = 0 \text{ or } v_{\text{esc}} = \sqrt{\frac{2GM}{R}}.$$

Thus, $v_{\text{esc}} = \sqrt{2}v$. $\qquad \Diamond$

P11.35 (a) What value of n_i is associated with the 94.96-nm spectral line in the Lyman hydrogen series?

(b) Could this wavelength be associated with the Paschen series or Balmer series?

Solution Our equation is $\dfrac{1}{\lambda} = R_H\left(\dfrac{1}{n_f^{\,2}} - \dfrac{1}{n_i^{\,2}}\right)$ where $R_H = 1.097 \times 10^7$ m^{-1} and for the

Lyman series, $n_f = 1$, and $n_i = 2$, 3, 4, ... Substituting the given values,

$$\frac{1}{94.96 \times 10^{-9} \text{ m}} = \left(1.097 \times 10^{-9} \text{ m}^{-1}\right)\left(1 - \frac{1}{n_i^{\,2}}\right).$$

(a) Solving for n_i, $n_i = 5$. ◊

(b) By Figure 11.20, spectral lines in the Balmer and Paschen series all have much longer wavelengths, since much smaller energy losses put the atom into energy levels 2 or 3. ◊

P11.39 A hydrogen atom is in its first excited state ($n = 2$). Using the Bohr theory of the atom,

(a) calculate the radius of the orbit.

(b) calculate the linear momentum of the electron.

(c) calculate the angular momentum of the electron.

(d) calculate the kinetic energy of the electron.

(e) calculate the potential energy of the system.

(f) calculate the total energy of the system.

Solution We note, during our calculations, that the nominal velocity of the electron is less than 1% of the speed of light; therefore, we do not need to use relativistic equations.

(a) By Bohr's model, $r_n = n^2 a_0 = 2^2(0.052\,9 \text{ nm}) = 2.12 \times 10^{-10}$ m. ◊

continued on next page

(b) Since $m_e vr = n\hbar$,

$$p = m_e v = \frac{n\hbar}{r} = \frac{2(1.054\,6 \times 10^{-34}\ \text{J·s})}{2.12 \times 10^{-10}\ \text{m}}$$

$$p = 9.97 \times 10^{-25}\ \text{kg·m/s} \qquad \Diamond$$

(c) $\vec{\mathbf{L}} = \vec{\mathbf{r}} \times \vec{\mathbf{p}}$ becomes $L = rp = n\hbar = 2.11 \times 10^{-34}\ \text{J·s}$ $\qquad \Diamond$

(d) Next, $v = \dfrac{p}{m_e} = \dfrac{9.97 \times 10^{-25}\ \text{kg·m/s}}{9.11 \times 10^{-31}\ \text{kg}} = 1.09 \times 10^6\ \text{m/s}.$

So $K = \dfrac{1}{2} m_e v^2 = \dfrac{(9.11 \times 10^{-31}\ \text{kg})(1.09 \times 10^6\ \text{m/s})^2}{2}$

$$= \frac{5.45 \times 10^{-19}\ \text{J}}{1.602 \times 10^{-19}\ \text{J/eV}} = 3.40\ \text{eV} \qquad \Diamond$$

(e) From Chapter 7, the electric potential energy is given by $\dfrac{k_e q_1 q_2}{r}$.

$$U = -\frac{k_e e^2}{r} = -\frac{(8.99 \times 10^9\ \text{N·m}^2/\text{C}^2)(1.602 \times 10^{-19}\ \text{C})^2}{2.12 \times 10^{-10}\ \text{m}}$$

and $U = -1.09 \times 10^{-18}\ \text{J} = -6.80\ \text{eV}.$ $\qquad \Diamond$

(f) Thus, $E = K + U = -5.45 \times 10^{-19}\ \text{J} = -3.40\ \text{eV}.$ $\qquad \Diamond$

The total energy ought to be equal to minus the kinetic energy: our answer confirms this, but we needed to use more than 3 significant figures for some constants. For an accurate answer from data with fewer significant figures, we could calculate the total energy first:

$$E = \frac{-k_e e^2}{2r} = \frac{-(8.99 \times 10^9\ \text{N·m}^2/\text{C}^2)(1.602 \times 10^{-19}\ \text{C})^2}{2(2.12 \times 10^{-10}\ \text{m})} = -5.45 \times 10^{-19}\ \text{J} \quad \Diamond$$

$$K = E - U = 5.45 \times 10^{-19}\ \text{J} = 3.40\ \text{eV} \qquad \Diamond$$

P11.53 Two hypothetical planets of masses m_1 and m_2 and radii r_1 and r_2, respectively, are nearly at rest when they are an infinite distance apart. Because of their gravitational attraction, they head toward each other on a collision course.

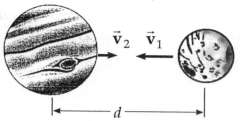

FIG. P11.53

(a) When their center-to-center separation is d, find expressions for the speed of each planet and for their relative speed.

(b) Find the kinetic energy of each planet just before they collide if $m_1 = 2.00 \times 10^{24}$ kg, $m_2 = 8.00 \times 10^{24}$ kg, $r_1 = 3.00 \times 10^6$ m, and $r_2 = 5.00 \times 10^6$ m. (**Note:** Both energy and momentum of the system are conserved.)

Solution We use both the energy version and the momentum version of the isolated system model.

(a) At infinite separation, $U = 0$; and at rest, $K = 0$. Since energy of the two-planet system is conserved, we have

$$0 = \frac{1}{2}m_1 v_1^2 + \frac{1}{2}m_2 v_2^2 - \frac{Gm_1 m_2}{d} \qquad (1)$$

The initial momentum of the system is zero and momentum is conserved. Therefore,

$$0 = m_1 v_1 - m_2 v_2 \qquad (2)$$

Combine Equations (1) and (2) to find

$$v_1 = m_2 \sqrt{\frac{2G}{d(m_1 + m_2)}} \text{ and } v_2 = m_1 \sqrt{\frac{2G}{d(m_1 + m_2)}}.$$

Relative velocity $v_r = v_1 - (-v_2) = \sqrt{\frac{2G(m_1 + m_2)}{d}}$. ◊

continued on next page

(b)   Substitute the given numerical values into the equation found for v_1 and v_2 in part (a) to find

$$v_1 = 1.03 \times 10^4 \text{ m/s and } v_2 = 2.58 \times 10^3 \text{ m/s}.$$

Therefore, $K_1 = \dfrac{1}{2} m_1 v_1^2 = 1.07 \times 10^{32} \text{ J}$; $K_2 = \dfrac{1}{2} m_2 v_2^2 = 2.67 \times 10^{31} \text{ J}$.   ◊

P11.59   Two stars of masses M and m, separated by a distance d, revolve in circular orbits about their center of mass (Fig. P11.59). Show that each star has a period given by

$$T^2 = \frac{4\pi^2}{G(M+m)} d^3$$

Proceed as follows: Apply Newton's second law to each star. Note that the center-of-mass condition requires that $Mr_2 = mr_1$, where $r_1 + r_2 = d$.

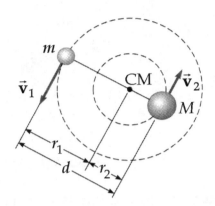

FIG. P11.59

Solution   For the star of mass M and orbital radius r_2, $\sum F = ma$ gives

$$\frac{GMm}{d^2} = \frac{Mv_2^2}{r_2} = \frac{M}{r_2}\left(\frac{2\pi r_2}{T}\right)^2.$$

For the star of mass m, $\sum F = ma$ gives $\dfrac{GMm}{d^2} = \dfrac{mv_1^2}{r_1} = \dfrac{m}{r_1}\left(\dfrac{2\pi r_1}{T}\right)^2$. Clearing fractions, we then obtain simultaneous equations: $GmT^2 = 4\pi^2 d^2 r_2$ and $GMT^2 = 4\pi^2 d^2 r_1$. Adding, we find

$$G(M+m)T^2 = 4\pi^2 d^2 (r_1 + r_2) = 4\pi^2 d^3$$

$$T^2 = \frac{4\pi^2 d^3}{G(M+m)} \qquad\qquad ◊$$

In a visual binary star system T, d, r_1, and r_2 can sometimes be measured, so the mass of each component can be computed. The equation derived represents a generalized version of Kepler's third law. It shows that the squared period is proportional to the cube of the orbit size.

P11.61 The positron is the antiparticle to the electron. It has the same mass and a positive electric charge of the same magnitude as that of the electron. Positronium is a hydrogen-like atom consisting of a positron and an electron revolving around each other. Using the Bohr model, find the allowed distances between the two particles and the allowed energies of the system.

Solution **Conceptualize:** Since we are told that positronium is like hydrogen, we might expect the allowed radii and energy levels to be about the same as for hydrogen:

$$r = a_0 n^2 = \left(5.29 \times 10^{-11} \text{ m}\right) n^2 \text{ and } E_n = \frac{-13.6 \text{ eV}}{n^2}.$$

Categorize: Similar to the textbook calculations for hydrogen, we can use the quantization of angular momentum of positronium to find the allowed radii and energy levels.

Analyze: Let r represent the distance between the electron and the positron. The two move in a circle of radius $\frac{r}{2}$ around their center of mass with opposite velocities. The total angular momentum is quantized according to

$$L_n = \frac{1}{2}mvr + \frac{1}{2}mvr = n\hbar \text{ where } n = 1, 2, 3, \ldots.$$

For each particle, $\sum F = ma$ expands to $\dfrac{k_e e^2}{r^2} = \dfrac{mv^2}{r/2}$.

We can eliminate $v = \dfrac{n\hbar}{mr}$ to find $\dfrac{k_e e^2}{r} = \dfrac{2mn^2\hbar^2}{m^2 r^2}$.

So the separation distances are $\quad r = \dfrac{2n^2\hbar^2}{mk_e e^2} = 2a_0 n^2 = \left(1.06 \times 10^{-10} \text{ m}\right) n^2.$ ◊

The orbital radii are $\dfrac{r}{2} = a_0 n^2$, the same as for the electron in hydrogen.

The energy can be calculated from $\quad E = K + U = \dfrac{1}{2}mv^2 + \dfrac{1}{2}mv^2 - \dfrac{k_e e^2}{r}.$

Since $mv^2 = \dfrac{k_e e^2}{2r}$,

$$E = \frac{k_e e^2}{2r} - \frac{k_e e^2}{r} = -\frac{k_e e^2}{2r} = \frac{-k_e e^2}{4a_0 n^2} = -\frac{6.80 \text{ eV}}{n^2}.$$ ◊

continued on next page

Finalize: It appears that the allowed separations for positronium are twice as large as for hydrogen, while the energy levels are half as big. One way to explain this is that in a hydrogen atom, the proton is much more massive than the electron, so the proton remains nearly stationary with essentially no kinetic energy. However, in positronium, the positron and electron have the same mass and therefore both have kinetic energy that separates them from each other and reduces the magnitude of their total energy compared with hydrogen.

Oscillatory Motion

NOTES FROM SELECTED CHAPTER SECTIONS

Section 12.1   Motion of a Particle Attached to a Spring

Simple harmonic motion of a mechanical system corresponds to the oscillation of an object between two points for an indefinite period of time, with no loss in mechanical energy.

The **mass-spring system** shown in Figure 12.1 is an example of a system that undergoes simple harmonic motion . The mass is assumed to move on a horizontal, frictionless surface. The point $x = 0$ is the equilibrium position of the mass; that is, the point where the mass would reside if left undisturbed. In this position, there is no horizontal force on the mass. When the mass is displaced a distance x from its equilibrium position, the spring produces a linear restoring force given by Hooke's law, $F = -kx$, where k is the force constant of the spring, and has SI units of N/m. The minus sign means that F is to the left when the displacement x is positive, whereas F is to the right when x is negative. *In other words, the direction of the force F is always towards the equilibrium position.*

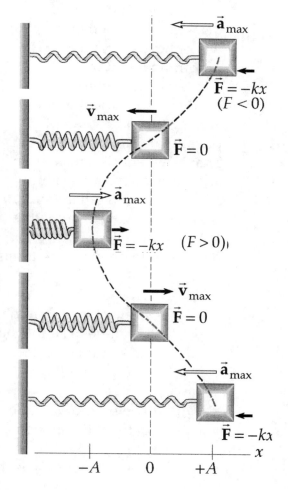

FIG. 12.1
Oscillating motion of a mass on the end of a spring.

Section 12.2   Mathematical Representation of Simple Harmonic Motion

The value of the phase constant ϕ depends on the initial displacement and initial velocity of the body. **(a)**

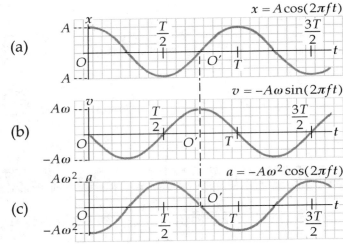

Figure 12.2 represents plots of the displacement, velocity, and acceleration versus time **(b)** assuming that at $t = 0$, $x_i = A$ and $v_i = 0$. In this case, one finds that $\phi = 0$. Note that the velocity is 90° out of phase **(c)** with the displacement. That is, v is zero when the displacement is a maximum, while the magnitude of v is a maximum when the displacement is zero. Furthermore, note that the acceleration is 180° out of phase with the displacement. That is, when x is a maximum and positive, a is a maximum, and negative. In other words, a is proportional to x, but in the opposite direction.

FIG. 12.2

Representation of (a) the displacement, (b) the velocity, and (c) the acceleration as a function of time for an object moving with simple harmonic motion.

It is useful to define a few terms relative to harmonic motion:

- The **amplitude**, A, is the maximum distance that an object moves away from its equilibrium position. In the absence of friction, an object will continue in simple harmonic motion. During each cycle, it will reach a maximum displacement on each side of the equilibrium position equal to the amplitude.

- The **period**, T, is the time it takes the object to execute one complete cycle of the motion.

- The **frequency**, f, is the number of cycles or vibrations per second.

Section 12.3   Energy Considerations in Simple Harmonic Motion

You should study carefully the comparison between the motion of the mass-spring system and that of the simple pendulum. In particular, notice that when the displacement is a maximum, the energy of the system is entirely potential energy; whereas, when the displacement is zero, the energy is entirely kinetic energy. For an arbitrary value of x, the energy is the sum of K and U.

Section 12.4   The Simple Pendulum

A simple pendulum consists of a mass m attached to a light string of length L as shown in Figure 12.3. When the angular displacement θ is small during the entire motion (less than about 15°), the pendulum approximates simple harmonic motion. In this case, the resultant force acting on the mass m equals the component of weight tangent to the circle, and has a magnitude $mg\sin\theta$. Since this force is always directed towards $\theta = 0$, it corresponds to a restoring force. For small θ, with θ measured in radians, we use the small angle approximation $\sin\theta \cong \theta$. In this approximation, the equation of motion reduces to Equation 12.23:

$$\frac{d^2\theta}{dt^2} = \frac{-g\theta}{L}.$$

This equation is identical in form to Eq. 12.5 for the mass-spring system, $\dfrac{d^2x}{dt^2} = -\omega^2 x$, which has the solution $x = A\cos(\omega t + \phi)$. The corresponding solution to Equation 12.23 is $\theta = \theta_i \cos(\omega t + \phi)$, where ω is given by Equation 12.24. The period of motion is given by Equation 12.25. In other words, the period depends only on the length of the pendulum and the acceleration of gravity. *The period does not depend on mass.*

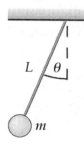

FIG. 12.3

Section 12.6   Damped Oscillations

Damped oscillations occur in realistic systems in which retarding forces such as friction are present. These forces will reduce the amplitudes of the oscillations with time, since mechanical energy is continually lost by the system. When the retarding force is assumed to be proportional to the velocity, but small compared to the restoring force, the system will still oscillate, but the amplitude will decrease exponentially with time.

EQUATIONS AND CONCEPTS

An object exhibits **simple harmonic motion** when the net force along the direction of motion is proportional to the displacement and in the opposite direction. The force constant, k, is always positive and has a value that corresponds to the relative stiffness of the spring. The negative sign means that the force exerted on the mass is always directed opposite the displacement.

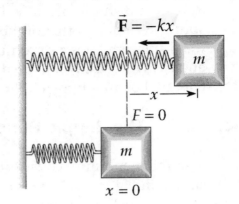

Hooke's law gives force exerted by a spring on a mass attached to the spring and displaced a distance x from the unstretched position. *The force is a restoring force, always directed toward the equilibrium position.*

$$F_s = -kx \tag{12.1}$$

The mathematical representation of a mass in simple harmonic motion along the x axis is a second order differential equation.

$$\frac{d^2x}{dt^2} = -\omega^2 x \tag{12.5}$$

$$\omega = \sqrt{\frac{k}{m}} \tag{12.9}$$

For an **ideal oscillator in harmonic motion** along the x axis, the following constants of the motion are found in expressions for position, velocity, and acceleration:

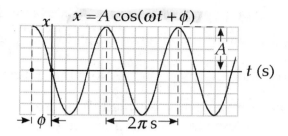

$$x = A\cos(\omega t + \phi)$$

Amplitude, A represents the maximum displacement along either positive or negative x.

For the displacement shown in the figure: $A = 4$ units

Angular frequency, ω has units of radians per second and is a measure of how rapidly the oscillations are occurring.

$$\omega = \frac{1 \text{ rev}}{2\pi \text{ s}} = \frac{2\pi \text{ rad}}{2\pi \text{ s}} = 1 \text{ rad/s}$$

Phase constant, ϕ is determined by the position and velocity of the oscillating particle when $t = 0$. The quantity $(\omega t + \phi)$ is called the phase.

$$\phi = \frac{+\pi}{2} \text{rad}$$

The **period of motion** T of a simple harmonic oscillator equals the time it takes the mass to complete one oscillation.

$$T = \frac{2\pi}{\omega} = 2\pi\sqrt{\frac{m}{k}} \tag{12.13}$$

The **frequency** of the motion, f, numerically equals the inverse of the period and represents the number of oscillations per unit time.

$$f = \frac{1}{T} = \frac{1}{2\pi}\sqrt{\frac{k}{m}} \tag{12.14}$$

The **position as a function of time** describes the displacement from equilibrium for a particle moving in simple harmonic motion along the x axis.

$$x(t) = A\cos(\omega t + \phi) \tag{12.6}$$

where $\omega^2 = \dfrac{k}{m}$ \tag{12.4}

The **velocity as a function of time** is found by taking the first derivative of x with respect to time.

$$v = \frac{dx}{dt} = -\omega A\sin(\omega t + \phi) \tag{12.15}$$

$$v_{\text{max}} = \omega A \tag{12.17}$$

The **acceleration as a function of time** is equal to the time derivative of the velocity (second derivative of the displacement). *Note that the acceleration (and hence the force) is always proportional to and opposite the displacement.*

$$a = \frac{d^2x}{dt^2} = -\omega^2 A\cos(\omega t + \phi) \tag{12.16}$$

$$a_{\text{max}} = \omega^2 A \tag{12.18}$$

The **total energy of a simple harmonic oscillator** is the sum of the kinetic energy $\left(\frac{1}{2}mv^2\right)$ and the potential energy $\left(\frac{1}{2}kx^2\right)$. Note that E remains constant since it is assumed there are no non-conservative forces acting on the system. The total energy of the simple harmonic oscillator is a constant of the motion and is proportional to the square of the amplitude.

$$E = \frac{1}{2}mv^2 + \frac{1}{2}kx^2$$

or

$$E = \frac{1}{2}kA^2 \tag{12.21}$$

The **velocity as a function of position** can be found from the principle of conservation of energy. *The magnitude of the velocity of an object in simple harmonic motion is a maximum at $x = 0$; the velocity is zero when the mass is at the points of maximum displacement $(x = \pm A)$.*

$$v = \pm\sqrt{\frac{k}{m}\left(A^2 - x^2\right)} \tag{12.22}$$

or

$$v = \pm\omega\sqrt{\left(A^2 - x^2\right)}$$

The **equation of motion for the simple pendulum** assumes a small displacement so that $\sin\theta \approx \theta$. *This small angle approximation is valid when the angle θ is measured in radians.*

$$\frac{d^2\theta}{dt^2} = -\frac{g}{L}\theta \tag{12.23}$$

The **period of a simple pendulum** depends only on the length of the supporting string and the value of the acceleration due to gravity.

$$\omega = \sqrt{\frac{g}{L}} \tag{12.24}$$

$$T = 2\pi\sqrt{\frac{L}{g}} \tag{12.25}$$

The **period of a physical pendulum,** for a small amplitude, depends on the moment of inertia, I, and the distance, d, between the pivot point and the center of mass. *This result can be used to measure the moment of inertia of a flat rigid body when the mass and the location of the center of mass are known.*

$$T = \frac{2\pi}{\omega} = 2\pi\sqrt{\frac{I}{mgd}}$$

SUGGESTIONS, SKILLS, AND STRATEGIES

Most of this chapter deals with simple harmonic motion, and the properties of the displacement expression

$$x(t) = A\cos(\omega t + \phi) \tag{12.6}$$

In order to obtain the velocity $v(t)$ and acceleration $a(t)$ of the system, one must be familiar with the derivative operation as applied to trigonometric functions. In particular, note that

$$\frac{d}{dt}\cos(\omega t + \phi) = -\omega\sin(\omega t + \phi) \qquad \frac{d}{dt}\sin(\omega t + \phi) = \omega\cos(\omega t + \phi).$$

Using these results, and $x(t)$ from Equation 12.6, we see that

$$v(t) = \frac{dx(t)}{dt} = -A\omega\sin(\omega t + \phi) \tag{12.15}$$

and

$$a(t) = \frac{dv(t)}{dt} = -A\omega^2\cos(\omega t + \phi) \tag{12.16}$$

By direct substitution, you should be able to show that Equation 12.1 represents a general solution to the equation of motion for the mass-spring system (a second-order homogeneous differential equation) given by

$$\frac{d^2x}{dt^2} + \frac{k}{m}x = 0 \text{ where } \omega = \sqrt{\frac{k}{m}}.$$

In treating the motion of the simple pendulum, we made use of the small angle approximation $\sin\theta \approx \theta$. This approximation enables us to reduce the equation of motion to that of the simple harmonic oscillator. The small angle approximation for $\sin\theta$ follows from inspecting the series expansion for $\sin\theta$, where θ is in radians:

$$\sin\theta = \theta - \frac{\theta^3}{3!} + \frac{\theta^5}{5!} - \ldots$$

For small values of θ, the higher order terms in θ^3, θ^5, ... are small compared to θ, so it follows that $\sin\theta \approx \theta$. The difference between $\sin\theta$ and θ is less than 1% for $0 < \theta < 15°$ (where $15° \approx 0.26$ rad).

The following table shows the similar structure, corresponding equations of motion, and the expression for the period of three simple oscillators.

Oscillating system	Equation of motion	Period
Harmonic oscillator along x axis	$\dfrac{d^2x}{dt^2} = -\left(\dfrac{k}{m}\right)x$	$T = 2\pi\sqrt{\dfrac{m}{k}}$
Simple Pendulum	$\dfrac{d^2\theta}{dt^2} = -\left(\dfrac{g}{L}\right)\theta$	$T = 2\pi\sqrt{\dfrac{L}{g}}$
Physical pendulum	$\dfrac{d^2\theta}{dt^2} = -\left(\dfrac{mgd}{I}\right)\theta$	$T = 2\pi\sqrt{\dfrac{I}{mgd}}$

REVIEW CHECKLIST

✓ Describe the general characteristics of simple harmonic motion, and the significance of the various parameters that appear in the expression for the displacement versus time, $x = A\cos(\omega t + \phi)$.

✓ Start with the general expression for the displacement versus time for a simple harmonic oscillator, and obtain expressions for the velocity and acceleration as functions of time.

✓ Determine the frequency, period, amplitude, phase constant, and position at a specified time of a simple harmonic oscillator given an equation for $x(t)$.

✓ Start with an expression for $x(t)$ for an oscillator and determine the maximum speed, maximum acceleration, and the total distance traveled in a specified time.

✓ Describe the phase relations among displacement, velocity, and acceleration for simple harmonic motion; and, given a curve of one of these variables verses time, sketch curves showing the time dependence of the other two.

✓ Calculate the mechanical energy, maximum velocity, and maximum acceleration of a mass-spring system given values for amplitude, spring constant , and mass.

ANSWERS TO SELECTED QUESTIONS

Q12.3 If the position of a particle varies as $x = -A \cos \omega t$, what is the phase constant in Equation 12.6? At what position is the particle at $t = 0$?

Answer Equation 12.6 says $x = A \cos(\omega t + \phi)$. Since negating a cosine wave is equivalent to changing the phase by 180°, we can say that

$$\phi = 180° \left(\frac{\pi \, \text{rad}}{180°} \right) = \pi \, \text{rad}.$$

At time $t = 0$, the particle is at a position of $x = -A$.

Q12.5 Determine whether the following quantities can be in the same direction for a simple harmonic oscillator:

(a) position and velocity

(b) velocity and acceleration

(c) position and acceleration.

Answer In a simple harmonic oscillator, the velocity follows the position by 1/4 of a cycle, and the acceleration follows the position by 1/2 of a cycle.

(a) Referring to Figure 12.2 of this study guide, it can be noted that there exist times when both the position and the velocity are positive, and therefore in the same direction.

(b) There also exist times when both the velocity and the acceleration are positive, and therefore in the same direction.

(c) On the other hand, when the position is positive, the acceleration is always negative, and therefore position and acceleration always have opposite signs.

Q12.13 Is it possible to have damped oscillations when a system is at resonance? Explain.

Answer Yes. At resonance, the amplitude of a damped oscillator will remain constant. If the system were not damped, the amplitude would increase without limit at resonance.

SOLUTIONS TO SELECTED PROBLEMS

P12.3   The position of a particle is given by the expression $x = (4.00 \text{ m})\cos(3.00\pi t + \pi)$, where x is in meters and t is in seconds.

  (a)   Determine the frequency and period of the motion.

  (b)   Determine the amplitude of the motion.

  (c)   Determine the phase constant.

  (d)   Determine the position of the particle at $t = 0.250$ s.

Solution   We use the particle in simple harmonic motion model. The particular position function $x = (4.00 \text{ m})\cos(3.00\pi t + \pi)$ and the general one, $x = A\cos(\omega t + \phi)$ have a specially powerful kind of equality called functional equality. They must give the same x value for all values of the variable t. This requires, then, that all parts be the same:

  (a) $\omega = 3.00\pi \text{ rad/s} = \left(\dfrac{2\pi \text{ rad}}{1 \text{ cycle}}\right) f$ or $f = 1.50 \text{ Hz}$,   ◊

  $T = \dfrac{1}{f} = 0.667 \text{ s}$ ◊

  (b) $A = 4.00 \text{ m}$ ◊

  (c) $\phi = \pi \text{ rad}$ ◊

  (d)   At $t = 0.250$ s, $x = (4.00 \text{ m})\cos(1.75\pi \text{ rad}) = (4.00 \text{ m})\cos(5.50 \text{ rad})$.

  Note that 5.50 rad is **not** 5.50°.

  Instead, $x = (4.00 \text{ m})\cos(5.50 \text{ rad}) = (4.00 \text{ m})\cos(315°) = 2.83 \text{ m}$.   ◊

P12.5   A particle moving along the x axis in simple harmonic motion starts from its equilibrium position, the origin, at $t = 0$ and moves to the right. The amplitude of its motion is 2.00 cm and the frequency is 1.50 Hz.

(a)   Show that the position of the particle is given by $x = (2.00 \text{ cm})\sin(3.00\pi t)$.

(b)   Determine the maximum speed and the earliest time $(t > 0)$ at which the particle has this speed

(c)   Determine the maximum acceleration and the earliest time $(t > 0)$ at which the particle has this acceleration, and

(d)   Determine the total distance traveled between $t = 0$ and $t = 1.00$ s.

Solution   (a)   At $t = 0$, $x = 0$ and v is positive (to the right). The sine function is zero and the cosine is positive at $\theta = 0$, so this situation corresponds to $x = A\sin\omega t$ and $v = v_i \cos\omega t$.

Since $f = 1.50$ Hz, $\omega = 2\pi f = 3\pi \text{s}^{-1}$.

Also, $A = 2.00$ cm

so that $x = (2.00 \text{ cm})\sin(3\pi t)$.   ◊

(b)   This is equivalent to writing $x = A\cos(\omega t + \phi)$

with $A = 2.00$ cm, $\omega = 3.00\pi \text{s}^{-1}$ and $\phi = -90° = -\dfrac{\pi}{2}$.

Note also that $T = \dfrac{1}{f} = 0.667$ s.

The velocity is $v = \dfrac{dx}{dt} = 2.00(3.00\pi)\cos(3.00\pi t)$ cm/s.

The maximum speed is $v_{max} = v_i = A\omega = 2.00(3.00\pi)$ cm/s $= 18.8$ cm/s.   ◊

This speed occurs at $t = 0$, when $\cos(3.00\pi t) = +1$

and next at $t = \dfrac{T}{2} = 0.333$ s, when $\cos\left[\left(3.00\pi \text{s}^{-1}\right)(0.333 \text{ s})\right] = -1$.   ◊

(c)   Again, $a = \dfrac{dv}{dt} = (-2.00 \text{ cm})(3.00\pi \text{s}^{-1})^2 \sin(3.00\pi t)$.

Its maximum value is $a_{max} = A\omega^2 = (2.00 \text{ cm})(3.00\pi \text{s}^{-1})^2 = 178$ cm/s^2

continued on next page

The acceleration has this positive value for the first time

at $t = \dfrac{3T}{4} = 0.500$ s   ◊

when $a = -(2.00 \text{ cm})(3.00\pi \text{s}^{-1})^2 \sin\left[(3.00\pi \text{s}^{-1})(0.500 \text{ s})\right] = 178 \text{ cm/s}^2$.   ◊

(d)   Since $A = 2.00$ cm, the particle will travel 8.00 cm in one period, $T = \dfrac{2}{3}$ s.

Hence, in $(1.00 \text{ s}) = \dfrac{3}{2}T$

the particle will travel $\dfrac{3}{2}(8.00 \text{ cm}) = 12.0 \text{ cm}$.   ◊

P12.9   A 7.00-kg object is hung from the bottom end of a vertical spring fastened to an overhead beam. The object is set into vertical oscillations with a period of 2.60 s. Find the force constant of the spring.

Solution   An object hanging from a vertical spring moves with simple harmonic motion just like an object moving without friction attached to a horizontal spring.

We are given the period. It is related to the angular frequency by $\dfrac{2\pi}{\omega} = T = 2.60$ s.

Solving for the angular frequency, $\omega = \dfrac{2\pi \text{ rad}}{2.60 \text{ s}} = 2.42 \text{ rad/s}$.

However, $\omega = \sqrt{\dfrac{k}{m}}$, so $k = \omega^2 m = (2.41 \text{ rad/s})^2 (7.00 \text{ kg})$.

Thus we solve for the force constant: $k = 40.9 \text{ kg/s}^2 = 40.9 \text{ N/m}$.   ◊

P12.11   A 0.500-kg object attached to a spring with a force constant 8.00 N/m vibrates in simple harmonic motion with an amplitude of 10.0 cm.

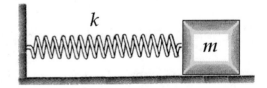

FIG. P12.11

(a)   Calculate the maximum value of its speed and acceleration.

(b)   Calculate the speed and acceleration when the object is 6.00 cm from the equilibrium position.

continued on next page

(c)　Calculate the time interval required for the object to move from $x = 0$ to $x = 8.00$ cm.

Solution　$\omega = \sqrt{\dfrac{k}{m}} = \sqrt{\dfrac{8.00 \text{ N/m}}{0.500 \text{ kg}}} = 4.00 \text{ s}^{-1}$. Therefore, position is given by

$$x = (10.0 \text{ cm}) \sin\left[\left(4.00 \text{ s}^{-1}\right)t\right].$$

(a)　From this we find that

$$v = \frac{dx}{dt} = (40.0 \text{ cm/s})\cos(4.00t)$$

$$v_{max} = 40.0 \text{ cm/s} \qquad \Diamond$$

$$a = \frac{dv}{dt} = -\left(160 \text{ cm/s}^2\right)\sin(4.00t)$$

$$a_{max} = 160 \text{ cm/s}^2 \qquad \Diamond$$

(b)　$t = \dfrac{1}{4.00}\sin^{-1}\left(\dfrac{x}{10.0 \text{ cm}}\right)$. When $x = 6.00$ cm, $t = 0.161$ s and we find that

$$v = (40.0 \text{ cm/s})\cos\left[\left(4.00 \text{ s}^{-1}\right)(0.161 \text{ s})\right] = 32.0 \text{ cm/s} \qquad \Diamond$$

$$a = -\left(160 \text{ cm/s}^2\right)\sin\left[\left(4.00 \text{ s}^{-1}\right)(0.161 \text{ s})\right] = -96.0 \text{ cm/s}^2 \qquad \Diamond$$

(c)　Using $t = \dfrac{1}{4}\sin^{-1}\left(\dfrac{x}{10.0 \text{ cm}}\right)$.

When $x = 0$, $t = 0$ and when $x = 8.00$ cm, $t = 0.232$ s.

Therefore $\Delta t = 0.232$ s. $\qquad \Diamond$

P12.13   An automobile having a mass of 1 000 kg is driven into a brick wall in a safety test. The bumper behaves like a spring of force constant 5.00×10^6 N/m, and compresses 3.16 cm as the car is brought to rest. What was the speed of the car before impact, assuming that no mechanical energy is lost during impact with the wall?

Solution   **Conceptualize:** If the bumper is only compressed 3 cm, the car is probably not permanently damaged, so v is most likely less than 10 mph ($\sim$5 m/s).

Categorize: Assuming no energy is lost during impact with the wall, the initial energy (kinetic) equals the final energy (elastic potential).

Analyze: Energy conservation gives

$$K_i = U_f \ \text{or} \ \frac{1}{2}mv^2 = \frac{1}{2}kx^2.$$

Solving for the velocity,

$$v = \sqrt{\frac{k}{m}}x = \left(3.16 \times 10^{-2} \ \text{m}\right)\sqrt{\frac{5.00 \times 10^6 \ \text{N/m}}{1\,000 \ \text{kg}}}.$$

Thus,

$$v = 2.23 \ \text{m/s}. \qquad\qquad \Diamond$$

Finalize: The speed is less than 5 m/s as predicted, so the answer seems reasonable. If the speed of the car were sufficient to compress the bumper beyond its elastic limit, then some of the initial kinetic energy would be lost to deforming the front of the car. In that case, some other procedure would have to be used to estimate the car's initial speed.

P12.21 A particle executes simple harmonic motion with an amplitude of 3.00 cm. At what position does its speed equal one half of its maximum speed?

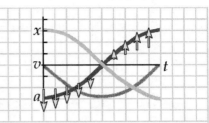

Solution **Conceptualize:** If we consider the speed of the particle along its path as shown in the graphical and tabular representations, we can see that the particle is at rest momentarily at one endpoint while being accelerated toward the middle by an elastic force that decreases as the particle approaches the equilibrium position. When it reaches the midpoint, the direction of acceleration changes so that the particle slows down until it stops momentarily at the opposite endpoint. From this analysis,

$\vec{x}$	3 cm	0 cm	−3 cm
$\vec{v}$	0	$-v_{max}$	0
$\vec{a}$	$-a_{max}$	0	a_{max}

FIG. P12.21

we can estimate that $v = \dfrac{v_{max}}{2}$ somewhere in the outer half of the travel (since this is the region where the speed is changing most rapidly): $1.50 < \pm x < 3.00$

Categorize: One way to analyze this problem is to start with the equation for linear SHM, take the first derivative with respect to time to find $v(t)$, and solve this equation for x when $v = \dfrac{v_{max}}{2}$.

Analyze: For linear SHM, $x = A\cos\omega t$

noting that the negative indicates direction, $\dfrac{dx}{dt} = v = -A\omega\sin\omega t$.

Since $A\omega$ is a constant, $v = \dfrac{v_{max}}{2}$ when $\quad \sin\omega t = \pm\dfrac{1}{2}$

or $\qquad \omega t = \sin^{-1}\left(\pm\dfrac{1}{2}\right) = \pm\dfrac{\pi}{6}$ or $\pm\dfrac{5}{6}\pi$.

Thus, in our displacement equation, $\qquad \cos\omega t = \cos\left(\pm\dfrac{\pi}{6}\right) = \dfrac{\sqrt{3}}{2}$

or $\qquad \cos\omega t = \cos\left(\pm\dfrac{5\pi}{6}\right) = -\dfrac{\sqrt{3}}{2}$.

continued on next page

Substituting $A = 3.00$ cm, $\qquad x = \pm\dfrac{A\sqrt{3}}{2} = \pm\dfrac{(3.00 \text{ cm})\sqrt{3}}{2}.$

Solving, $\qquad x = \pm 2.60$ cm. $\qquad\qquad\qquad\qquad$ ◊

We could get the same answer from $\qquad v = \pm\omega\sqrt{A^2 - x^2}.$

Finalize: The calculated position is in the outer half of the travel as predicted, and is in fact very close to the endpoints. This means that the speed of the particle changes remarkably little until the particle reaches the ends of its travel, where it experiences the maximum restoring force of the spring, which is proportional to x.

P12.23 A simple pendulum has a mass of 0.250 kg and a length of 1.00 m. It is displaced through an angle of 15.0° and then released.

(a) What is the maximum speed?

(b) What is the maximum angular acceleration?

(c) What is the maximum restoring force?

Solve this problem once by using the simple harmonic motion model for the motion of the pendulum, and then solve the problem more precisely by using more general principles.

Solution **METHOD ONE:**

Since 15.0° is small enough that (in radians) $\sin\theta \approx \theta$ within 1%, we may model the motion as simple harmonic motion. The constant angular frequency characterizing the motion is

$$\omega = \sqrt{\frac{g}{L}} = \sqrt{\frac{9.80 \text{ m/s}^2}{1.00 \text{ m}}} = 3.13 \text{ rad/s}$$

The amplitude as a distance is $A = L\theta = (1.00 \text{ m})(0.262 \text{ rad}) = 0.262 \text{ m}.$

(a) The maximum linear speed is $v_{\max} = \omega A = (3.13 \text{ s}^{-1})(0.262 \text{ m}) = 0.820 \text{ m/s}.$ ◊

continued on next page

(b)   Similarly, $a_{max} = \omega^2 A = (3.13\ 1/s)^2\ (0.262\ m) = 2.57\ m/s^2$. This implies maximum angular acceleration

$$\alpha = \frac{a}{r} = \frac{2.57\ m/s^2}{1.00\ m} = 2.57\ rad/s^2. \qquad \lozenge$$

(c) $\sum F = ma = (0.250\ kg)(2.57\ m/s^2) = 0.641\ N \qquad \lozenge$

METHOD TWO:

We may work out slightly more precise answers by using the energy version of the isolated system model and the particle under a net force model. At release, the pendulum has height above its equilibrium position.

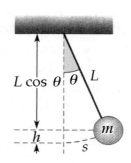

$$h = L - L\cos 15.0° = (1.00\ m)(1 - \cos 15.0°) = 0.034\ 1\ m$$

(a)   The energy of the pendulum-Earth system is conserved as the pendulum swings down:

$$(K + U)_{top} = (K + U)_{bottom}$$

$$0 + mgh = \frac{1}{2}mv_{max}^2 + 0$$

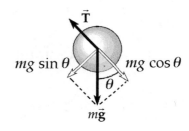

$$v_{max} = \sqrt{2gh} = \sqrt{2(9.80\ m/s^2)(0.034\ 1\ m)}$$

FIG. P12.23

$$v_{max} = 0.817\ m/s \qquad \lozenge$$

(c)   The restoring force at release is

$$mg\sin 15.0° = (0.250\ kg)(9.80\ m/s^2)(\sin 15.0°) = 0.634\ N \qquad \lozenge$$

(b)   This produces linear acceleration $a = \dfrac{\sum F}{m} = \dfrac{0.634N}{0.250\ kg} = 2.54\ m/s^2$

and angular acceleration $\alpha = \dfrac{a}{r} = \dfrac{2.54\ m/s^2}{1.00\ m} = 2.54\ rad/s^2. \qquad \lozenge$

P12.25 A particle of mass m slides without a friction inside hemispherical bowl of radius R. Show that if it starts from rest with a small displacement from equilibrium, the particle moves in simple harmonic motion with an angular frequency equal to that of a simple pendulum of length R (that is,

$$\omega = \sqrt{\frac{g}{R}}).$$

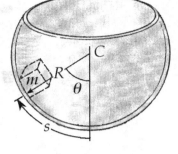

Solution Locate the center of curvature C of the bowl. We can measure the excursion of the object from equilibrium by the angle θ between the radial line to C and the vertical. The distance the object moves from equilibrium is $s = R\theta$.

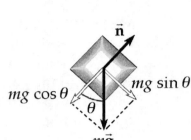

FIG. P12.25

$\sum F_s = ma$ becomes $\qquad -mg\sin\theta = m\dfrac{d^2s}{dt^2}$.

For small angles $\qquad\qquad \sin\theta \approx \theta$

so by substitution, $\qquad -mg\theta = m\dfrac{d^2s}{dt^2}; \; -mg\dfrac{s}{R} = m\dfrac{d^2s}{dt^2}$.

Isolating the derivative, $\qquad \dfrac{d^2s}{dt^2} = -\left(\dfrac{g}{R}\right)s$.

By the form of this equation, we can see that the acceleration is proportional to the position and in the opposite direction, so we have SHM. $\qquad\qquad \lozenge$

We identify its angular frequency by comparing our equation to Eq. 12.5:
$$\dfrac{d^2x}{dt^2} = -\omega^2 x.$$

Now x and s both measure position, so $\omega^2 = \dfrac{g}{R}$ and $\omega = \sqrt{\dfrac{g}{R}}$. $\qquad\qquad \lozenge$

P12.27   A physical pendulum in the form of a planar body moves in simple harmonic motion with a frequency of 0.450 Hz. The pendulum has a mass of 2.20 kg and the pivot is located 0.350 m from the center of mass. Determine the moment of inertia of the pendulum about the pivot point.

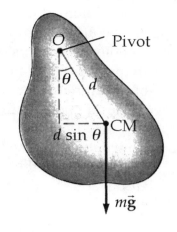

Solution $f = 0.450$ Hz; $d = 0.350$ m; $m = 2.20$ kg.

Using Eq. 12.27, $T = 2\pi\sqrt{\dfrac{I}{mgd}}$ and $T^2 = \dfrac{4\pi^2 I}{mgd}$.

FIG. P12.27

$$I = \frac{T^2 mgd}{4\pi^2} = \left(\frac{1}{f}\right)^2 \frac{mgd}{4\pi^2} = \frac{(2.20 \text{ kg})(9.80 \text{ m/s}^2)(0.350 \text{ m})}{(0.450 \text{ s}^{-1})^2 (4\pi^2)} = 0.944 \text{ kg} \cdot \text{m}^2$$ ◊

P12.41   A large block P executes horizontal simple harmonic motion as it slides across a frictionless surface with a frequency $f = 1.50$ Hz. Block B rests on it, as shown in Figure P12.41, and the coefficient of static friction between the two is $\mu_s = 0.600$. What maximum amplitude of oscillation can the system have if block B is not to slip?

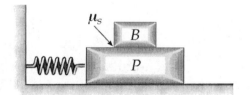

Solution   If the block B does not slip, it undergoes simple harmonic motion (SHM) with the same amplitude and frequency as those of P, and with its acceleration caused by the static friction force exerted on it by P. Think of the block when it is just ready to slip at a turning point in its motion: $\sum F = ma$ becomes

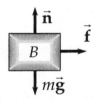

FIG. P12.41

$$f_{max} = \mu_s n = \mu_s mg = ma_{max} = mA\omega^2.$$

Then $A = \dfrac{\mu_s g}{\omega^2} = \dfrac{0.600(9.80 \text{ m/s}^2)}{[2\pi(1.50/\text{s})]^2} = 6.62$ cm.   ◊

P12.47   A pendulum of length L and mass M has a spring of
force constant k connected to it at a distance h below
its point of suspension (Fig. P12.47). Find the
frequency of vibration of the system for small values
of the amplitude (small θ). Assume the vertical
suspension of length L is rigid, but ignore its mass.

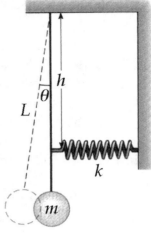

Solution   **Conceptualize:** The frequency of vibration should
be greater than that of a simple pendulum since the
spring adds on additional restoring force:

$$f > \frac{1}{2\pi}\sqrt{\frac{g}{L}}.$$

Categorize: We can find the frequency of oscillation
from the angular frequency, ω, which is found in
the equation for angular SHM:

$$\frac{d^2\theta}{dt^2} = -\omega^2\theta.$$

The angular acceleration can be found from
analyzing the torques acting on the pendulum.

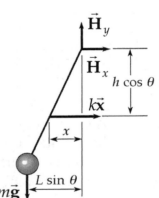

Analyze: For the pendulum (see sketch)

FIG. P12.47

$$\sum \tau = I\alpha \quad \text{and} \quad \frac{d^2\theta}{dt^2} = -\alpha.$$

The negative sign appears because positive θ is measured clockwise in the
picture. We take torque around the point of suspension:

$$\sum \tau = MgL\sin\theta + kxh\cos\theta = I\alpha.$$

For small amplitude vibrations, use the approximations:

$$\sin\theta \approx \theta,, \qquad \cos\theta \approx 1, \qquad \text{and } x = h\tan\theta \approx h\theta.$$

Therefore, with $I = mL^2$, $\dfrac{d^2\theta}{dt^2} = -\left(\dfrac{MgL + kh^2}{I}\right)\theta = -\left(\dfrac{MgL + kh^2}{ML^2}\right)\theta.$

continued on next page

This is of the SHM form

$$\frac{d^2\theta}{dt^2} = -\omega^2\theta$$

with angular frequency,

$$\omega = \sqrt{\frac{MgL + kh^2}{ML^2}} = 2\pi f.$$

The ordinary frequency is

$$f = \frac{\omega}{2\pi} = \frac{1}{2\pi L}\sqrt{\frac{MgL + kh^2}{M}}.$$ ◊

Finalize: The frequency is greater than for a simple pendulum as we expected. In fact, the additional contribution inside the square root looks like the frequency of a mass on a spring scaled by $\dfrac{h}{L}$ since the spring is connected to the rod and not directly to the mass. So we can think of the solution as:

$$f^2 = \frac{1}{4\pi^2}\left(\frac{MgL + kh^2}{ML^2}\right) = f_{\text{pendulum}}^2 + \frac{h^2}{L^2} f_{\text{spring}}^2.$$

P12.53   A ball of mass m is connected to two rubber bands of length L, each under tension T, as in Figure P12.53. The ball is displaced by a small distance y perpendicular to the length of the rubber bands.

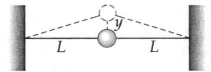

FIG. P12.53

(a)   Assuming the tension does not change, show that the restoring force is $-\dfrac{2T}{L}y$.

(b)   Assuming the tension does not change, show that the system exhibits simple harmonic motion with an angular frequency $\omega = \sqrt{\dfrac{2T}{mL}}$.

Solution   (a) $\sum \vec{F} = (-2T\sin\theta)\hat{j}$ where $\theta = \tan^{-1}\left(\dfrac{y}{L}\right)$.

Since for a small displacement, $\sin\theta \approx \tan\theta = \dfrac{y}{L}$

and the resultant force is $\sum \vec{F} = \left(-\dfrac{2Ty}{L}\right)\hat{j}.$ ◊

continued on next page

(b)   Since there is a restoring force that is proportional to the position, it causes the system to move with simple harmonic motion like a block-spring system.

Thus, $\sum \vec{\mathbf{F}} = -k\vec{\mathbf{x}}$ becomes $\qquad \sum F = -\left(\dfrac{2T}{L}\right)y$.

Therefore, $\qquad\qquad\qquad \omega = \sqrt{\dfrac{k}{m}} = \sqrt{\dfrac{2T}{mL}}$. $\qquad\qquad\qquad\qquad \diamond$

Raising the string tension increases the frequency and increasing the mass decreases the frequency, both through proportionalities to the square root.

Chapter 13

Mechanical Waves

NOTES FROM SELECTED CHAPTER SECTIONS

Section 13.1   Propagation of a Disturbance

A mechanical wave requires:

- a source of disturbance

- a medium that can be disturbed

- a physical mechanism through which particles of the medium can influence one another.

The wave function $y(x,t)$ represents a one-dimensional wave pulse and gives the y coordinate of any point P located at position x at any time t.

A transverse wave is one in which the particles of the disturbed medium oscillate back and forth along a direction perpendicular to the direction of the wave velocity.

A longitudinal wave is one in which the particles of the medium undergo a displacement (oscillate back and forth) along a direction parallel to the direction of the wave velocity (direction along which the pulse travels).

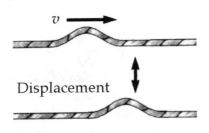
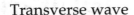

Displacement

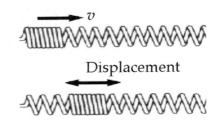

Displacement

Transverse wave   Longitudinal wave

A one-dimensional traveling wave can be described mathematically by its wave function $y(x,t) = f(x \mp vt)$. If the wave is assumed to be traveling along the x direction, then $y(x, t)$ represents the y coordinate at position x on the string at any time, t.

In the expression for the wave function, the negative sign describes a wave pulse traveling toward the right and the positive sign describes the wave function for a pulse traveling toward the left.

In the case of a wave on a string:

- **at a fixed value of** x, the wave function represents the y coordinate of a particular point as a function of time

- **at a given value** t, the wave function defines a curve showing the shape of the wave pulse at a given time.

Section 13.2   The Wave Model

There are three important parameters in characterizing sinusoidal waves:

- **Wavelength** is the minimum distance between two points that are the same distance from their equilibrium positions and are moving in the same direction (such a pair of points are said to be in phase).

- **Frequency** is the rate at which the disturbance of the elastic medium repeats itself.

- **Speed** is the rate at which the wave propagates through the medium.

A sinusoidal wave is one whose shape is sinusoidal at every instant of time. Two waves which differ in phase are shown in the adjacent figure.

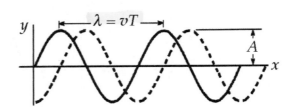

The **wave amplitude**, A, is the maximum possible value of the displacement of a particle of the medium away from its equilibrium position.

The **period of the wave** is the time required for the disturbance (or pulse) to travel along the direction of propagation a distance equal to one wavelength. The period is also the time required for any point in the medium to complete one complete cycle in its harmonic motion about its equilibrium point.

Section 13.4   The Speed of Transverse Waves on Strings

The **speed of mechanical waves** depends only on the physical properties of the medium through which the disturbance travels. In the case of waves on a string, the speed depends on the tension in the string and the mass per unit length (linear mass density).

Section 13.5   Reflection and Transmission of Waves

The following general rules apply to reflected waves: When a wave pulse travels **from** medium A to medium B and $v_A > v_B$ (that is, when B is more dense than A), the reflected part of the pulse is inverted upon reflection. When a wave pulse travels **from** medium A to medium B and $v_A < v_B$ (A is more dense than B), the reflected pulse is not inverted.

Section 13.6   Rate of Energy Transfer by Sinusoidal Waves on Strings

As waves propagate through a medium, they transport energy; this occurs without any net transfer of matter. The power transmitted by any sinusoidal wave is proportional to the square of the angular frequency and to the square of the amplitude.

Section 13.7   Sound Waves

Sound waves are longitudinal waves. The particles of the medium through which longitudinal waves travel, oscillate back and forth parallel to the direction of prorogation of the waves. This is in contrast to a transverse wave, in which the vibrations of the medium are perpendicular to the direction of travel of the wave.

Sound waves may be described as displacement waves or as pressure waves. The pressure is proportional to the amplitude of the displacement; however, the pressure wave is 90° out of phase with the displacement wave.

Section 13.8   The Doppler Effect

In general, a Doppler effect is experienced whenever there is relative motion between source and observer. When the source and observer are moving toward each other, the frequency heard by the observer is higher than the frequency of the source. When the source and observer are moving away from each other, the observer hears a frequency lower than the source frequency.

EQUATIONS AND CONCEPTS

The **wave function** $y(x, t)$ represents the y coordinate of an element or small segment of a medium as a wave pulse travels along the x direction. The value of y depends on two variables (position and time) and is read "y as a function of x and t".) *When t has a fixed value, $y = y(x)$ and defines the shape of the wave pulse at an instant in time.*

$$y(x,t) = f(x - vt) \quad \text{(13.1a)}$$
pulse traveling right

$$y(x,t) = f(x + vt) \quad \text{(13.1b)}$$
pulse traveling left

The displacement repeats itself when x is increased by an integral multiple of λ and the wave moves to the right a distance of vt in a time t.

$$y = A\sin\left(\frac{2\pi}{\lambda}(x - vt)\right) \quad \text{(13.4)}$$

The **wave speed** can be expressed in terms of the wavelength and period. *A traveling wave moves a distance $x = \lambda$ in a time $t = T$.*

$$v = \frac{\lambda}{T} \quad \text{(13.5)}$$

It is convenient to define three additional characteristic wave quantities:

wave number k,

$$k \equiv \frac{2\pi}{\lambda} \quad \text{(13.7)}$$

angular frequency ω,

$$\omega \equiv \frac{2\pi}{T} = 2\pi f \quad \text{(13.8)}$$

frequency f.

$$f = \frac{1}{T}$$

The **wave function** can be written in a more compact form in terms of the parameters defined above. If the transverse displacement is not zero at $x = 0$ and $t = 0$, it is necessary to include a phase constant, ϕ.

$$y = A\sin(kx - \omega t) \tag{13.9}$$

$$y = A\sin(kx - \omega t + \phi) \tag{13.12}$$

The **wave speed** v or phase velocity can also be expressed in alternative forms.

$$v = \frac{\omega}{k} \tag{13.10}$$

$$v = \lambda f \tag{13.11}$$

The **transverse speed** v_y and **transverse acceleration** a_y of a point on a sinusoidal wave on a string are out of phase by $\dfrac{\pi}{2}$ radians. *Do not confuse transverse speed with wave speed;* v_y *and* a_y *describe the motion of an element of the string that moves perpendicular to the direction of propagation of the wave.* Note expressions for the maximum values of transverse speed and acceleration.

$$v_y = -\omega A\cos(kx - \omega t) \tag{13.13}$$

$$v_{y,\max} = \omega A \tag{13.15}$$

$$a_y = -\omega^2 A\sin(kx - \omega t) \tag{13.14}$$

$$a_{y,\max} = \omega^2 A \tag{13.16}$$

The **linear wave equation** is satisfied by any wave function having the form $y = f(x \pm vt)$.

$$\frac{\partial^2 y}{\partial x^2} = \frac{1}{v^2}\frac{\partial^2 y}{\partial t^2} \tag{13.19}$$

For a **transverse pulse in a stretched string**, the wave speed depends on the tension in the string T and the linear density μ of the string (mass per unit length).

$$v = \sqrt{\frac{T}{\mu}} \qquad T = v^2\mu \tag{13.20}$$

The **power** (rate of transfer of energy) transmitted by any harmonic wave is proportional to the square of the frequency and the square of the amplitude, where μ is the mass per unit length of the string.

$$\mathcal{P} = \frac{1}{2}\mu\omega^2 A^2 v \tag{13.23}$$

The **displacement from equilibrium** of an element of medium in which a harmonic sound wave is propagating has a sinusoidal variation in time. The amplitude (maximum displacement) is $s_{\max}$. *The displacement is parallel to the direction of the propagation of the wave (longitudinal wave).*

$$s(x, t) = s_{\max}\sin(kx - \omega t) \tag{13.24}$$

The **pressure variations** in a medium conducting a sound wave vary harmonically in time and are out of phase with the displacements by $\dfrac{\pi}{2}$ radians (compare Equations 13.24 and 13.25).

$$\Delta P = \Delta P_{max} \cos(kx - \omega t) \qquad (13.25)$$

The **pressure amplitude** is proportional to the displacement amplitude.

$$\Delta P_{max} = \rho v \omega s_{max} \qquad (13.26)$$

The **Doppler effect** (apparent shift in frequency) is observed whenever there is relative motion between a source and an observer. *Equation 13.30 is a general Doppler shift expression; it applies to relative motion of source and observer toward or away from each other with correct use of algebraic signs. In Equation 13.30, v_0 and v_s are measured relative to the medium in which the sound travels.*

$$f' = f\left(\frac{v + v_0}{v - v_s}\right) \qquad (13.30)$$

where

v = speed of sound in a medium,
v_0 = velocity of the observer, and
v_s = velocity of the source.

For both v_0 and v_s, a positive value is used when the corresponding object moves toward the other object; negative values are used when the object moves away from the other.

REVIEW CHECKLIST

✓   Recognize whether or not a given function is a possible description of a traveling wave.

✓   Express a given harmonic wave function in several alternative forms involving different combinations of the wave parameters: wavelength, period, phase velocity, wave number, angular frequency, and harmonic frequency.

✓   Given a specific wave function for a harmonic wave, obtain values for the characteristic wave parameters: A, ω, k, λ, f, and ϕ.

✓   Calculate the rate at which energy is transported by harmonic waves in a string.

✓   Describe the harmonic displacement and pressure variation as functions of time and position for a harmonic sound wave. Relate the displacement amplitude to the pressure amplitude for a harmonic sound wave and calculate the wave intensity from each of these parameters.

✓   Describe the various situations under which a Doppler shifted frequency is produced. Note that a Doppler shift is observed as long as there is a relative motion between the observer and the source.

✓   Equation 13.30 is the generalized equation for the Doppler effect:

$$f' = \left(\frac{v + v_0}{v - v_s} \right) f$$

Where:
f = source frequency
f' = observed frequency
v = speed of sound in medium

v_0 = speed of observer
v_s = speed of source

The most likely error in using Equation 13.30 is using the incorrect algebraic sign for the velocity of either the observer or the source.
When the relative motion of source and observer is:

Either or both toward the other: **Enter v_s and v_0 with + signs.**
Either or both away from the other: **Enter v_s and v_0 with – signs.**

Remember in Equation 13.30 stated above, the algebraic signs (the plus sign in the numerator and the minus sign in the denominator) are part of the structure of the equation. **These signs remain and correct signs, stated above, must be entered along with the values of v_0 and v_s.** You should confirm the correct use of signs in the general Doppler equation for the cases of relative motion in the table below.

Observer	Source	Equation	Remark
O →	S	$f' = f\left(\dfrac{v + v_0}{v}\right)$	Observer moving toward stationary source
← O	S	$f' = f\left(\dfrac{v - v_0}{v}\right)$	Observer moving away from stationary source
O	← S	$f' = f\left(\dfrac{v}{v - v_s}\right)$	Source moving toward stationary observer
O	S →	$f' = f\left(\dfrac{v}{v + v_s}\right)$	Source moving away from stationary observer
O →	S →	$f' = f\left(\dfrac{v + v_0}{v + v_s}\right)$	Observer following moving source
← O	← S	$f' = f\left(\dfrac{v - v_0}{v - v_s}\right)$	Source following moving observer
← O	S →	$f' = f\left(\dfrac{v - v_0}{v + v_s}\right)$	Observer and source both moving away from each other
O →	← S	$f' = f\left(\dfrac{v + v_0}{v - v_s}\right)$	Observer and source moving toward each other
O	S	$f' = f$	Observer and source both stationary

EXAMPLES OF DOPPLER EFFECT WITH OBSERVER/SOURCE IN MOTION

ANSWERS TO SELECTED QUESTIONS

Q13.1 How would you create a longitudinal wave in a stretched spring? Would it be possible to create a transverse wave in a spring?

Answer A longitudinal wave can be set up in a stretched spring by compressing the coils in a small region, and releasing the compressed region. The disturbance will proceed to propagate as a longitudinal pulse. It is quite possible to set up a transverse wave in a spring, simply by displacing the spring in a direction perpendicular to its length.

Q13.7   If you stretch a rubber hose and pluck it, you can observe a pulse traveling up and down the hose. What happens to the speed of the pulse if you stretch the hose more tightly? What happens to the speed if you fill the hose with water?

Answer   If you stretch the hose tighter, you increase the tension, and increase the speed of the wave. If the hose elongates at all, then you decrease the linear density, which also increases the speed of the wave. If you fill it with water, you increase the linear density of the hose, and decrease the speed of the wave.

Q13.9   A vibrating source generates a sinusoidal wave on a string under constant tension. If the power delivered to the string is doubled, by what factor does the amplitude change? Does the wave speed change under these circumstances?

Answer   Power is always proportional to the square of the amplitude, so if power doubles, amplitude increases by $\sqrt{2}$ times, a factor of 1.41. The wave speed does not depend on amplitude, but stays constant.

Q13.11   If an alarm clock is placed in a good vacuum and then activated, no sound is heard. Explain.

Answer   We assume that a perfect vacuum surrounds the clock. The sound waves require a medium for them to travel to your ear. The hammer on the alarm will strike the bell, and the vibration will spread as sound waves through the body of the clock. If a bone of your skull were in contact with the clock, you would hear the bell. However, in the absence of a surrounding medium like air or water, no sound can be radiated away.

What happens to the sound energy? Here is the answer: As the sound wave travels through the steel and plastic, traversing joints and going around corners, its energy is converted into additional internal energy, raising the temperature of the materials. After the sound has died away, the clock will glow very slightly brighter in the infrared spectrum. For a larger-scale example of the same effect: Colossal storms raging on the Sun are deathly still for us.

Q13.15   How can an object move with respect to an observer so that the sound from it is not shifted in frequency?

Answer   For the sound from a source not to shift in frequency, the radial velocity of the source relative to the observer must be zero; that is, the source must not be moving toward or away from the observer.

This can happen if the source and observer are not moving at all; if they have equal velocities relative to the medium; or, it can happen if the source moves around the observer in a circular pattern of constant radius. Even if the source accelerates along the circle, decelerates, or stops, the frequency heard will equal the frequency emitted by the source.

SOLUTIONS TO SELECTED PROBLEMS

P13.1 At $t = 0$, a transverse pulse in a wire is described by the function

$$y = \frac{6}{x^2 + 3}$$

where x and y are in meters. Write the function $y(x, t)$ that describes this pulse if it is traveling in the positive x direction with a speed of 4.50 m/s.

Solution $y(x, t)$ must be a function of both x and t, but must become $y = \frac{6}{x^2 + 3}$ when $t = 0$.

To guarantee the same form, we substitute the term $x = x' + ut$, and solve for y:

$$y = \frac{6}{(x' + ut)^2 + 3}$$

Note that as t increases, x' must decrease by $u\Delta t$; aside from that, the equation remains the same. We first define that at $t = 0$, $x = 0$. So

$$x' + u(0) = x = 0 \tag{1}$$

In order to cause the wave to appear to move to the right, we need to force our reference point (x) to the left. Therefore, 1 second later, at $t = 1$, the pulse has moved 4.50 m in the $+x$ direction, but x moves 4.50 m in the $-x$ direction.

$$x' + u(1 \text{ s}) = x = -4.50 \text{ m} \tag{2}$$

Subtracting equations (1) and (2), $u = -4.50$ m/s, and our new equation is:

$$y(x, t) = \frac{6}{(x - 4.50t)^2 + 3} \qquad \Diamond$$

In general, we can cause any waveform to move along the x axis at a velocity v_x by substituting $(x - v_x t)$ for x. The same principle applies to motion in other directions.

P13.3 A sinusoidal wave is traveling along a rope. The oscillator that generates the wave completes 40.0 vibrations in 30.0 s. Also, a given maximum travels 425 cm along the rope in 10.0 s. What is the wavelength?

Solution $f = \dfrac{40.0 \text{ waves}}{30.0 \text{ s}} = 1.33 \text{ s}^{-1}$ and $v = \dfrac{425 \text{ cm}}{10.0 \text{ s}} = 42.5 \text{ cm/s}.$

Since $v = \lambda f$, $\qquad\qquad \lambda = \dfrac{v}{f} = \dfrac{42.5 \text{ cm/s}}{1.33 \text{ s}^{-1}} = 0.319 \text{ m}.$ ◊

P13.5 The wave function for a traveling wave on a taut string is (in SI units)

$$y(x,\ t) = (0.350 \text{ m}) \sin\left(10\pi t - 3\pi x + \frac{\pi}{4}\right)$$

(a) What are the speed and direction of travel of the wave?

(b) What is the vertical displacement of the string at $t = 0$, $x = 0.100$ m?

(c) What are the wavelength and frequency of the wave?

(d) What is the maximum magnitude of the transverse speed of an element of the string?

Solution We use the traveling wave model. We compare the given equation with $y = A\sin(kx - \omega t + \phi)$. We note that $\sin(\theta) = -\sin(-\theta) = \sin(-\theta + \pi)$ and find that $k = 3\pi$ rad/m and $\omega = 10\pi$ rad/s.

(a) The speed and direction of the wave can be defined in terms of the velocity:

$$\vec{v} = f\lambda\hat{\mathbf{i}} = \frac{\omega}{k}\hat{\mathbf{i}} = \frac{10\pi \text{ rad/s}}{3\pi \text{ rad/m}}\hat{\mathbf{i}} = 3.33\hat{\mathbf{i}} \text{ m/s}$$ ◊

(b) Substituting $t = 0$ and $x = 0.100$ m,

$$y = (0.350 \text{ m})\sin(-0.300\pi + 0.250\pi) = -0.0548 \text{ m} = -5.48 \text{ cm}$$ ◊

Note that when you take the sine of a quantity with no units, it is not in degrees, but in radians.

continued on next page

(c) $\lambda = \dfrac{2\pi\,\text{rad}}{k} = \dfrac{2\pi\,\text{rad}}{3\pi\,\text{rad/m}} = 0.667$ m   ◊

and $f = \dfrac{\omega}{2\pi\,\text{rad}} = \dfrac{10\pi\,\text{rad/s}}{2\pi\,\text{rad}} = 5.00$ Hz   ◊

(d) $v_y = \dfrac{\partial y}{\partial t} = (0.350\text{ m})(10\pi\text{ rad/s})\cos\left(10\pi t - 3\pi x + \dfrac{\pi}{4}\right)$. The maximum occurs

when the cosine term is 1:

$$v_{y,\text{max}} = (10\pi\text{ rad/s})(0.350\text{ m/s}) = 11.0\text{ m/s}\ .$$ ◊

Note the large difference between the maximum particle speed and the wave speed found in part (a).

P13.9   (a)   Write the expression for y as a function of x and t for a sinusoidal wave traveling along a rope in the **negative** x direction with the following characteristics: $A = 8.00$ cm, $\lambda = 80.0$ cm, $f = 3.00$ Hz, and $y(0,t) = 0$ at $t = 0$.

(b)   Write the expression for y as a function of x and t for the wave in (a) assuming that $y(x,\,0) = 0$ at the point $x = 10.0$ cm.

Solution The amplitude is

$$A = y_{\text{max}} = 8.00\text{ cm} = 0.0800\text{ m}$$

$$k = \dfrac{2\pi}{\lambda} = \dfrac{2\pi}{0.800\text{ m}} = 7.85\text{ m}^{-1}$$

$$\omega = 2\pi f = 2\pi(3.00\text{ s}^{-1}) = 6.00\pi\text{ rad/s}$$

(a)   Since $\phi = 0$, $y = A\sin(kx + \omega t + \phi)$

becomes $y = (0.080\,0\text{ m})\sin(7.85x + 6.00\pi t)$.   ◊

(b)   In general, $y = (0.080\,0\text{ m})\sin(7.85x + 6.00\pi t + \phi)$.

If $y(x,\,0) = 0$ at $x = 0.100$ m, $0 = (0.080\,0\text{ m})\sin(0.785 + \phi)$

and $\phi = -0.785$ rad.

Therefore, $y = (0.080\,0\text{ m})\sin(7.85x + 6.00\pi t - 0.785)$.   ◊

P13.17 A 30.0-m steel wire and a 20.0-m copper wire, both with 1.00-mm diameters, are connected end to end and stretched to a tension of 150 N. How long does it take a transverse wave to travel the entire length of the two wires?

Solution The total time of travel is the sum of the two times.

In each wire, $t = \dfrac{L}{v} = L\sqrt{\dfrac{\mu}{T}}$ where $\mu = \dfrac{m}{L} = \rho\dfrac{V}{L} = \rho A = \dfrac{\pi\rho d^2}{4}$

so $t = L\sqrt{\dfrac{\pi\rho d^2}{4T}}$.

For copper, $t_1 = (20.0 \text{ m})\sqrt{\dfrac{\pi\left(8\,920 \text{ kg/m}^3\right)\left(0.001\,00 \text{ m}\right)^2}{4\left(150 \text{ kg}\cdot\text{m/s}^2\right)}} = 0.137 \text{ s}$

For steel, $t_2 = (30.0 \text{ m})\sqrt{\dfrac{\pi\left(7\,860 \text{ kg/m}^3\right)\left(0.001\,00 \text{ m}\right)^2}{4\left(150 \text{ kg}\cdot\text{m/s}^2\right)}} = 0.192 \text{ s}$

The total time is $(0.137 \text{ s}) + (0.192 \text{ s}) = 0.329 \text{ s}$. ◊

P13.21 Sinusoidal waves 5.00 cm in amplitude are to be transmitted along a string that has a linear mass density of 4.00×10^{-2} kg/m. If the source can deliver a maximum power of 300 W and the string is under a tension of 100 N, what is the highest vibrational frequency at which the source can operate?

Solution The wave speed is $v = \sqrt{\dfrac{T}{\mu}} = \sqrt{\dfrac{100 \text{ N}}{4.00\times10^{-2} \text{ kg/m}}} = 50.0 \text{ m/s}$.

From $\mathcal{P} = \dfrac{1}{2}\mu\omega^2 A^2 v$, $\omega^2 = \dfrac{2\mathcal{P}}{\mu A^2 v} = \dfrac{2(300 \text{ N}\cdot\text{m/s})}{\left(4.00\times10^{-2} \text{ kg/m}\right)\left(5.00\times10^{-2} \text{ m}\right)^2(50.0 \text{ m/s})}$.

Solving, $\omega = 346.4$ rad/s and $f = \dfrac{\omega}{2\pi} = 55.1 \text{ Hz}$. ◊

P13.23   Suppose that you hear a clap of thunder 16.2 s after seeing the associated lightning stroke. The speed of sound waves in air is 343 m/s, and the speed of light in air is 3.00×10^8 m/s. How far are you from the lightning stroke?

Solution   **Conceptualize:** There is a common rule of thumb that lightning is about a mile away for every 5 seconds of delay between the flash and thunder (or ~ 3 s/km). Therefore, this lightning strike is about 3 miles (~ 5 km) away.

Categorize: The distance can be found from the speed of sound and the elapsed time. The time for the light to travel to the observer will be much less than the sound delay, so the speed of light can be taken as ∞.

Analyze: Assuming that the speed of sound is constant through the air between the lightning strike and the observer,

$$v_s = \frac{d}{\Delta t}$$

or $d = v_s \Delta t = (343 \text{ m/s})(16.2 \text{ s}) = 5.56 \text{ km}.$ ◊

Finalize: Our calculated answer is consistent with our initial estimate, but we should check the validity of our assumption that the speed of light could be ignored. The time delay for the light is

$$t_{\text{light}} = \frac{d}{c_{\text{air}}} = \frac{5\,560 \text{ m}}{3.00 \times 10^8 \text{ m/s}} = 1.85 \times 10^{-5} \text{ s}$$

and $\Delta t = t_{\text{sound}} - t_{\text{light}} = 16.2 \text{ s} - 1.85 \times 10^{-5} \text{ s} \approx 16.2 \text{ s}.$

Since the travel time for the light is much smaller than the uncertainty in the time of 16.2 s, t_{light} can be ignored without affecting the distance calculation. However, our assumption of a constant speed of sound in air is not precisely valid due to local variations in air temperature during a storm. We must assume that the given speed of sound in air is an accurate **average** value for the conditions described.

P13.29   An experimenter wishes to generate in air a sound wave that has a displacement amplitude of 5.50×10^{-6} m. The pressure amplitude is to be limited to 0.840 N/m². What is the minimum wavelength the sound wave can have?

Solution   We are given $s_{max} = 5.50 \times 10^{-6}$ m and $\Delta P_{max} = 0.840$ Pa.

The pressure amplitude is $\Delta P_{max} = \rho v \omega s_{max} = \rho v \left(\dfrac{2\pi v}{\lambda} \right) s_{max}$

or $\lambda_{min} = \dfrac{2\pi \rho v^2 s_{max}}{\Delta P_{max}}$: $\lambda_{min} = \dfrac{2\pi \left(1.20 \text{ kg/m}^3 \right) \left(343 \text{ m/s} \right)^2 \left(5.50 \times 10^{-6} \text{ m} \right)}{0.840 \text{ Pa}} = 5.81 \text{ m}$ ◊

P13.31   Write an expression that describes the pressure variation as a function of position and time for a sinusoidal sound wave in air if $\lambda = 0.100$ m and $\Delta P_{max} = 0.200$ N/m².

Solution   We write the pressure variation as $\Delta P = \Delta P_{max} \sin(kx - \omega t)$.

Noting that $k = \dfrac{2\pi}{\lambda}$, $\qquad\qquad k = \dfrac{2\pi \text{ rad}}{0.100 \text{ m}} = 62.8 \text{ rad/m}$.

Likewise, $\omega = \dfrac{2\pi v}{\lambda}$, so $\omega = \dfrac{(2\pi \text{ rad})(343 \text{ m/s})}{0.100 \text{ m}} = 2.16 \times 10^4 \text{ rad/s}$.

We now can create our equation: $\Delta P = \left(0.200 \text{ N/m}^2 \right) \sin(62.8x - 21\,600t)$ ◊

where x is in meters and t is in seconds.

P13.35   Standing at a crosswalk, you hear a frequency of 560 Hz from the siren of an approaching ambulance. After the ambulance passes, the observed frequency of the siren is 480 Hz. Determine the ambulance's speed from these observations.

Solution   **Conceptualize:** We can assume that an ambulance with its siren on is in a hurry to get somewhere, and is probably traveling between 20 and 100 mi/h (~10 m/s to 50 m/s), depending on the driving conditions.

Categorize: We can use the equation for the Doppler effect to find the speed of the car.

Analyze : Let v_s represent the magnitude of the velocity of the ambulance.

continued on next page

Approaching car: $f' = \left(\dfrac{v}{v - v_s}\right)f$. Departing car: $f'' = \left(\dfrac{v}{v + v_s}\right)f$

where $f' = 560$ Hz and $f'' = 480$ Hz.

Solving the two equations above for f and setting them equal gives:

$$f'\left(1 - \frac{v_s}{v}\right) = f''\left(1 + \frac{v_s}{v}\right) \qquad \text{or} \qquad f' - f'' = (f' + f'')\left(\frac{v_s}{v}\right)$$

so the speed of the source is

$$v_s = \frac{v(f' - f'')}{f' + f''} = \frac{(343 \text{ m/s})(560 \text{ Hz} - 480 \text{ Hz})}{560 \text{ Hz} + 480 \text{ Hz}} = 26.4 \text{ m/s}. \qquad \lozenge$$

Finalize: This seems like a reasonable speed (about 50 mi/h) for an ambulance, unless the street is crowded or the car is traveling on an open highway. Of course, this problem is valid if the siren emits a single tone. If the siren warbles up and down in frequency, the maximum frequency in each cycle of vibrato could be measured and the problem solved in the same way.

P13.37 A tuning fork vibrating at 512 Hz falls from rest and accelerates at 9.80 m/s^2. How far below the point of release is the tuning fork when waves of frequency of 485 Hz reach the release point? Take the speed of sound in air to be 340 m/s.

Solution In order to solve this problem, we must first determine how fast the tuning fork is falling when its frequency is 485 Hz. The tuning fork (source) is moving **away** from a stationary listener.

Therefore, we use the equation $f' = \left(\dfrac{v}{v + v_s}\right)f$,

$$485 \text{ Hz} = (512 \text{ Hz})\frac{340 \text{ m/s}}{340 \text{ m/s} + v_{\text{fall}}}.$$

Solving, $v_{\text{fall}} = (340 \text{ m/s})\left(\dfrac{512 \text{ Hz}}{485 \text{ Hz}} - 1\right) = 18.93 \text{ m/s}.$

For the tuning fork we use the particle under constant acceleration model.

continued on next page

From the kinematic equation $v_2^2 = v_1^2 + 2as$,

we calculate that $s = \dfrac{v_2^2}{2a} = \dfrac{(18.93 \text{ m/s})^2}{2(9.80 \text{ m/s}^2)} = 18.28$ m.

Since $v = at$, $t = \dfrac{v_{\text{fall}}}{a} = \dfrac{18.93 \text{ m/s}}{9.80 \text{ m/s}^2} = 1.931$ s.

At this moment, the fork would appear to ring at 485 Hz to an observer just above the fork. However, it takes some additional time for the waves to reach the point of release. From the traveling wave model,

$$\Delta t = \frac{s}{v} = \frac{18.28 \text{ m}}{340 \text{ m/s}} = 0.053\ 8 \text{ s.}$$

Over the total time $t + \Delta t$, the fork falls a distance

$$s_{\text{total}} = \frac{1}{2}a(t + \Delta t)^2 = \frac{1}{2}(9.80 \text{ m/s}^2)(1.985 \text{ s})^2 = 19.3 \text{ m.} \qquad \Diamond$$

P13.47 A rope of total mass m and length L is suspended vertically. Show that a transverse wave pulse travels the length of the rope in a time interval $\Delta t = 2\sqrt{\dfrac{L}{g}}$.

(*Suggestion:* First find an expression for the wave speed at any point a distance x from the lower end by considering the tension in the rope as resulting from the weight of the segment below that point.)

Solution We define $x = 0$ at the bottom of the rope and $x = L$ at the top of the rope. The tension in the rope at any point is the weight of the rope below that point. We can thus write the tension in the rope at each point x as $T = \mu x g$, where μ is the mass per unit length of the rope.
The speed of the wave pulse at each point along the rope's length is therefore

$$v = \sqrt{\frac{F}{\mu}} \qquad \text{or} \qquad v = \sqrt{gx}.$$

But at each point x, the wave phase progresses at a rate of $v = \dfrac{dx}{dt}$.

continued on next page

So we can then substitute for v, and generate the differential equation:

$$\frac{dx}{dt} = \sqrt{gx} \qquad \text{or} \qquad dt = \frac{dx}{\sqrt{gx}}.$$

Integrating both sides, $\Delta t = \dfrac{1}{\sqrt{g}} \displaystyle\int_0^L \frac{dx}{\sqrt{x}} = \frac{\left[2\sqrt{x}\right]_0^L}{\sqrt{g}} = 2\sqrt{\frac{L}{g}}.$

We observe that the derived equation is dimensionally correct. In terms of units,

$$\left[2\sqrt{\frac{L}{g}}\right] = \sqrt{\frac{\text{m}}{\text{m}/\text{s}^2}} = \sqrt{\text{s}^2} = \text{s}.$$

A longer rope imposes a longer travel time on the pulse, but the rope must be 4 times longer for the travel time to be twice as large. The tension depends on the mass of the rope, but the pulse transit time does not depend on the mass. It divided out. The phenomenon described in the problem can be observed experimentally. Take a long bead chain. Let it hang at rest as you sight down along it, by holding one end between fingers of your left hand just below your eye. Tap near the top of the chain with your right hand. You can see a pulse go down, reflect at the bottom with no phase reversal, and come back up. Can you verify the equation derived here? Can you observe that the speed changes?

Chapter 14

Superposition and Standing Waves

Section 14.1 The Principle of Superposition

The superposition principle is that when two or more waves move in the same linear medium, the net displacement of the medium at any point (the resultant wave) equals the algebraic sum of the displacements of all the waves. If the individual waves are sinusoidal and of equal frequency, the resultant wave function is also sinusoidal and has the same frequency and same wavelength as the individual waves.

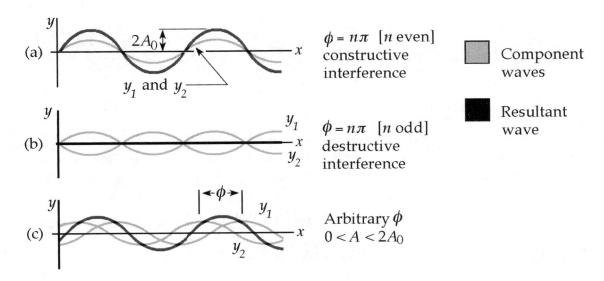

FIG. 14.1 Superposition of two waves with the same amplitude and frequency, with a phase difference ϕ of (a) $n\pi$ (n even), (b) $n\pi$ (n odd), and (c) arbitrary ϕ. The gray curves represent y_1 and y_2; the dark curves represent $y = y_1 + y_2$.

Figure 14.1 shows the resultant wave of two traveling sinusoidal waves for

(a) $\phi = 0$, 2π, 4π, ... corresponding to constructive interference,

(b) $\phi = \pi$, 3π, 5π, ... corresponding to destructive interference, and

(c) $0 < \phi < \pi$ for which the resultant amplitude has a value between 0 and $2A$.

Section 14.4   Standing Waves in Strings

Standing waves can be set up in a string by a continuous superposition of waves incident on and reflected from the ends of the string. The string has a number of natural patterns or frequencies of vibration, called normal modes or harmonics. Each harmonic has a characteristic frequency. The lowest of these frequencies is called the fundamental frequency, which together with the higher frequencies form a harmonic series. Figure 14.2 is a schematic representation of the first three normal modes of vibration of string fixed at both ends.

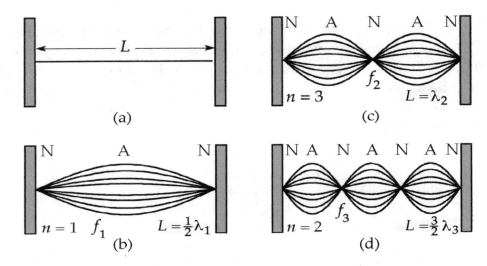

FIG. 14.2 Schematic representation of standing waves on a stretched string of length L, where the envelope represents many successive vibrations. The points of zero displacement are called nodes; the points of maximum displacement are called antinodes. The pattern shown in (b) represents the first harmonic or fundamental frequency.

Section 14.5   Standing Waves in Air Columns

Standing waves are produced in strings by interfering transverse waves. Sound sources can be used to produce longitudinal standing waves in air columns. The phase relationship between incident and reflected waves depends on whether or not the reflecting end of the air column is open or closed. This gives rise to two sets of possible standing wave conditions.

The first three natural modes of vibration for (a) an open pipe and (b) a closed pipe are shown in Figure 14.3.

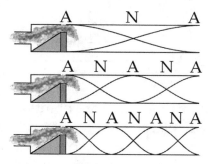

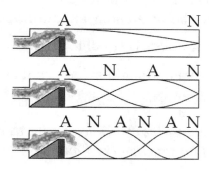

FIG. 14.3 (a) Natural modes of vibration in a hollow pipe open at each end. All harmonics are present.

FIG. 14.3 (b) Natural modes of vibration for air in a hollow pipe closed at one end. Only odd harmonic are present.

EQUATIONS AND CONCEPTS

The **resultant wave function** which is the resultant of two traveling sinusoidal waves having the same direction, frequency, and amplitude is also sinusoidal and has the same frequency and wavelength as the individual waves. The amplitude of the resultant wave depends on the phase difference ϕ between the two individual waves.

$$y = \left[2A\cos\frac{\phi}{2} \right] \sin\left(kx - \omega t + \frac{\phi}{2} \right) \qquad (14.1)$$

$$y_{\max} = 2A\cos\frac{\phi}{2}$$

A **phase difference** can arise between two waves generated by the same source and arriving at a common point after having traveled along paths of unequal path length.

$$\Delta r = \frac{\phi}{2\pi}\lambda \qquad (14.2)$$

ϕ is the phase difference
Δr is the path difference

A **standing wave** (an oscillation pattern) can be produced in a string due to the interference of two sinusoidal waves with equal amplitude and frequency traveling in opposite directions.

$$y = (2A\sin kx)\cos\omega t \qquad (14.3)$$

The **amplitude** ($2A \sin kx$ in Eq. 14.3) at any point along a standing wave on a string is a function of its position x along the string. The distance between adjacent antinodes or between adjacent nodes is equal to $\dfrac{\lambda}{2}$; and the distance between a node and an adjacent antinodes is equal to $\dfrac{\lambda}{4}$. *The "standing" wave is not a "traveling" wave because the expression for the wave does not contain the function* ($kx - \omega t$).

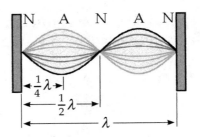

antinodes (A) are points of maximum displacement

nodes (N) are points of zero displacement.

Normal modes of oscillation (a series of natural patterns of vibration) can be excited in a string of length L fixed at each end. Each mode corresponds to a quantized frequency and wavelength. *The frequencies are integral multiples of a fundamental frequency (when $n = 1$) and can be expressed in terms of wave speed and string length or in terms of string tension and linear mass-density.*

$$\lambda_n = \frac{2L}{n} \qquad n = 1,\ 2,\ 3,\ \ldots \qquad (14.6)$$

$$f_n = \frac{n}{2L} v \qquad n = 1,\ 2,\ 3,\ \ldots \qquad (14.7)$$

$$f_n = \frac{n}{2L} \sqrt{\frac{T}{\mu}} \qquad n = 1,\ 2,\ 3,\ \ldots \qquad (14.8)$$

In an **"open"** pipe (open at both ends), the natural frequencies of oscillation form a harmonic series that includes all integral multiples of the fundamental frequency. *All harmonics are possible.*

$$f_n = n \frac{v}{2L} \qquad n = 1,\ 2,\ 3,\ \ldots \qquad (14.10)$$

(open pipe)

In a **"closed"** pipe (closed at one end), the natural frequencies of oscillation form a harmonic series that includes only the odd integral multiples of the fundamental frequency. *Only the odd harmonics are possible.*

$$f_n = n \frac{v}{4L} \qquad n = 1,\ 3,\ 5,\ \ldots \qquad (14.11)$$

(closed pipe)

Beats are produced by the superposition of two waves of equal amplitude but slightly different frequencies. *The resultant wave is simple harmonic with an effective frequency equal to the average of the two individual wave frequencies.*

$$y = \left[2A \cos 2\pi \left(\frac{f_1 - f_2}{2} \right) t \right] \cos 2\pi \left(\frac{f_1 + f_2}{2} \right) t$$

$$(14.12)$$

The **amplitude of the resultant wave** described by Equation 14.12 above is time dependent. *There are two maxima in each period of the resultant wave; and each occurrence of maximum amplitude results in a "beat".* The **beat frequency** f_b equals the difference in the frequencies of the two individual waves.

$$A_{x=0} = 2A\cos\left(2\pi\left(\frac{f_1 - f_2}{2}\right)t\right) \qquad (14.13)$$

$$f_b = |f_1 - f_2| \qquad (14.14)$$

Any **complex periodic wave form**, with a period T, can be represented by the combination of sinusoidal waves which form a harmonic series (combination of fundamental and various harmonics). The lowest frequency is $f_1 = \dfrac{1}{T}$ and higher frequencies are $f_n = nf_1$. A_n and B_n represent the amplitudes of the various harmonics.

$$y(t) = \sum_n \left(A_n \sin 2\pi f_n t + B_n \cos 2\pi f_n t\right)$$

$$(14.15)$$

A Fourier series

REVIEW CHECKLIST

✓ Write out the wave function that represents the superposition of the two sinusoidal waves of equal amplitude and frequency traveling in opposite directions in the same medium.

✓ Given an expression for the wave function, identify the angular frequency and maximum amplitude. Determine the values of x that correspond to nodal and antinodal points of a standing wave.

✓ Calculate the normal mode frequencies for a string under tension and for open and closed air columns.

✓ Describe the time dependent amplitude and determine the effective frequency of vibration when two waves of slightly different frequency interfere. Also, calculate the expected beat frequency for this situation.

ANSWERS TO SELECTED QUESTIONS

Q14.1 Does the phenomenon of wave interference apply only to sinusoidal waves?

FIG. Q14.1

Answer No. Any waves moving in the same medium can interfere with each other. For example, two pulses moving in opposite directions on a stretched string interfere when they meet each other.

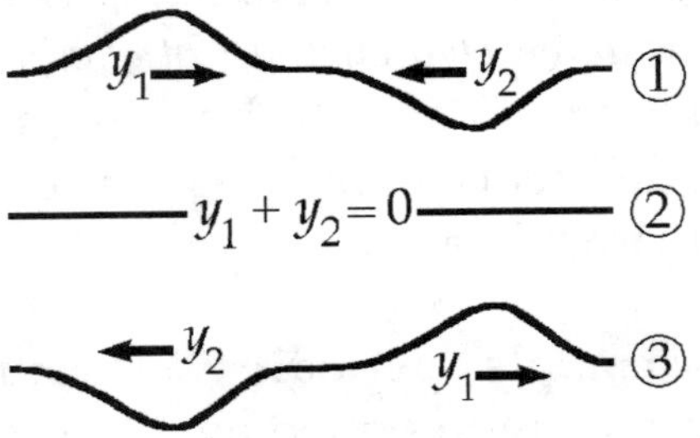

Q14.3 When two waves interfere, can the amplitude of the resultant wave be greater than either of the two original waves? If so, under what conditions can that happen?

Answer They can, wherever the two waves are enough in phase that their displacements will add to create a total displacement greater than the amplitude of either of the two original waves.

When two one-dimensional waves of the same amplitude interfere, these conditions are satisfied whenever the absolute value of the phase difference between the two waves is less than 120°.

Q14.5 When two waves interfere constructively or destructively, is there any gain or loss in energy? Explain.

Answer No. The energy may be transformed into other forms of energy. For example, when two pulses traveling on a stretched string in opposite directions overlap, and one is inverted, some potential energy is transferred to kinetic energy when they overlap. In fact, if they have equal amplitudes in opposite directions, they completely cancel each other at one point. In this case, all of the energy is transverse kinetic energy when the resultant amplitude is zero.

Q14.9 Why does a vibrating guitar string sound louder when placed on the instrument than it would if allowed to vibrate in the air while off the instrument?

Answer A vibrating string is not able to set very much air into motion when vibrated alone. Thus it will not be very loud. If it is placed on the instrument, however, the string's vibration sets the sounding board of the guitar into vibration. A vibrating piece of wood is able to move a lot of air, and the note is louder.

Q14.13 An airplane mechanic notices that the sound from a twin-engine aircraft rapidly varies in loudness when both engines are running. What could be causing this variation from loud to soft?

Answer Apparently the two engines are emitting sounds having frequencies which differ only by a very small amount from each other. This results in a beat frequency, causing the variation from loud to soft, and back again.

SOLUTIONS TO SELECTED PROBLEMS

P14.5 Two traveling sinusoidal waves are described by the wave functions y

$$y_1 = (5.00 \text{ m}) \sin[\pi(4.00x - 1\,200t)]$$

and

$$y_2 = (5.00 \text{ m}) \sin[\pi(4.00x - 1\,200t - 0.250)]$$

where x, y_1, and y_2 are in meters and t is in seconds.

(a) What is the amplitude of the resultant wave?

(b) What is the frequency of the resultant wave?

Solution We can represent the waves symbolically as

$$y_1 = A_0 \sin(kx - \omega t) \quad \text{and} \quad y_2 = A_0 \sin(kx - \omega t - \phi)$$

with $A_0 = 5.00$ m, $\omega = 1\,200\pi\,\text{s}^{-1}$, and $\phi = 0.250\pi$.

According to the principle of superposition, the resultant wave function has the form

$$y = y_1 + y_2 = 2A_0 \cos\left(\frac{\phi}{2}\right) \sin\left(kx - \omega t - \frac{\phi}{2}\right).$$

(a) with amplitude $A = 2A_0 \cos\left(\dfrac{\phi}{2}\right) = 2(5.00) \cos\left(\dfrac{\pi}{8.00}\right) = 9.24$ m ◊

(b) and frequency $f = \dfrac{\omega}{2\pi} = \dfrac{1\,200\pi}{2\pi} = 600$ Hz ◊

P14.13   Two speakers are driven in phase by a common oscillator at 800 Hz and face each other at a distance of 1.25 m. Locate the points along a line joining the two speakers where relative minima of sound pressure amplitude would be expected. (Use $v = 343$ m/s.)

$$\longleftarrow x \ \text{m} \longrightarrow | \longleftarrow (1.25 - x) \ \text{m} \longrightarrow$$

FIG. P14.13

Solution   The wavelength is $\lambda = \dfrac{v}{f} = \dfrac{343 \ \text{m/s}}{800 \ \text{Hz}} = 0.429$ m.

The two waves moving in opposite directions along the line between the two speakers will add to produce a standing wave with this distance between nodes:

$$\text{distance N to N} = \frac{\lambda}{2} = 0.214 \ \text{m}.$$

Because the speakers vibrate in phase, air compressions from each will simultaneously reach the point halfway between the speakers, to produce an antinode of pressure here. A node of pressure will be located at this distance on either side of the midpoint:

$$\text{distance N to A} = \frac{\lambda}{4} = 0.107 \ \text{m}.$$

Therefore nodes of sound pressure will appear at these distances from either speaker:

$$\frac{1}{2}(1.25 \ \text{m}) + 0.107 \ \text{m} = 0.732 \ \text{m}, \ \frac{1}{2}(1.25 \ \text{m}) - 0.107 \ \text{m} = 0.518 \ \text{m}.$$

The standing wave contains a chain of equally-spaced nodes at distances from either speaker of

0.732 m + 0.214 m = 0.947 m,   0.947 m + 0.214 m = 1.16 m

and also at 0.518 m − 0.214 m = 0.303 m,   0.303 m − 0.214 m = 0.089 1 m.

The standing wave exists only along the line segment between the speakers. No nodes or antinodes appear at distances greater than 1.25 m or less than 0, because waves add to give standing wave only if they are traveling in opposite directions and not in the same direction. In order, the distances from either speaker to the nodes of pressure between the speakers are at 0.089 1 m, 0.303 m, 0.518 m, 0.732 m, 0.947 m, and 1.16 m.   ◊

P14.15   Two sinusoidal waves combining in a medium are described by the wave functions

$$y_1 = (3.0 \text{ cm})\sin \pi(x + 0.60t) \text{ and } y_2 = (3.0 \text{ cm})\sin \pi(x - 0.60t)$$

where x is in centimeters and t is in seconds.

(a)   Determine the **maximum** transverse position of an element of the medium at $x = 0.250$ cm.

(b)   Determine the **maximum** transverse position of an element of the medium at $x = 0.500$ cm.

(c)   Determine the **maximum** transverse position of an element of the medium at $x = 1.50$ cm.

(d)   Find the three smallest values of x corresponding to antinodes.

Solution   According to the waves in interference model, we add y_1 and y_2 using the trigonometry identity

$$\sin(\alpha + \beta) = \sin\alpha\cos\beta + \cos\alpha\sin\beta.$$

We get $y = y_1 + y_2 = (6.0 \text{ cm})\sin(\pi x)\cos(0.60\pi t)$.

Since $\cos(0) = 1$, we can find the maximum value of y by setting $t = 0$:

$$y_{max}(x) = y_1 + y_2 = (6.0 \text{ cm})\sin(\pi x).$$

(a)   At $x = 0.250$ cm, $y_{max} = (6.0 \text{ cm})\sin(0.250\pi) = 4.24$ cm   ◊

(b)   At $x = 0.500$ cm, $y_{max} = (6.0 \text{ cm})\sin(0.500\pi) = 6.00$ cm   ◊

(c)   At $x = 1.50$ cm, $y_{max} = |(6.0 \text{ cm})\sin(1.50\pi)| = +6.00$ cm   ◊

(d)   The antinodes occur when $x = \dfrac{n\lambda}{4}$ $(n = 1, 3, 5,...)$.

But $k = \dfrac{2\pi}{\lambda} = \pi$, so $\lambda = 2.00$ cm and

$$x_1 = \frac{\lambda}{4} = \frac{2.00 \text{ cm}}{4} = 0.500 \text{ cm}$$ ◊

$$x_2 = \frac{3\lambda}{4} = \frac{3(2.00 \text{ cm})}{4} = 1.50 \text{ cm}$$ ◊

$$x_3 = \frac{5\lambda}{4} = \frac{5(2.00 \text{ cm})}{4} = 2.50 \text{ cm}$$ ◊

P14.17   Find the fundamental frequency and the next three frequencies that could cause standing wave patterns on a string that is 30.0 m long, has a mass per length of 9.00×10^{-3} kg/m, and is stretched to a tension of 20.0 N.

Solution   **Conceptualize:** The string described in the problem is very long, loose, and somewhat heavy, so it should have a very low fundamental frequency, maybe only a few vibrations per second.

Categorize: The tension and linear density of the string can be used to find the wave speed, which can then be used along with the required wavelength to find the fundamental frequency.

Analyze: The wave speed is

$$v = \sqrt{\frac{T}{\mu}} = \sqrt{\frac{20.0 \text{ N}}{9.00 \times 10^{-3} \text{ kg/m}}} = 47.1 \text{ m/s}.$$

For a vibrating string of length L fixed at both ends, the wavelength of the fundamental is $\lambda = 2L = 60.0$ m; and the frequency is

$$f_1 = \frac{v}{\lambda} = \frac{v}{2L} = \frac{47.1 \text{ m/s}}{60.0 \text{ m}} = 0.786 \text{ Hz}. \qquad \Diamond$$

The next three harmonics are

$$f_2 = 2f_1 = 1.57 \text{ Hz}, \ f_3 = 3f_1 = 2.36 \text{ Hz}, \text{ and } f_4 = 4f_1 = 3.14 \text{ Hz}. \qquad \Diamond$$

Finalize: The fundamental frequency is even lower than expected, less than 1 Hz. You could watch the string vibrating. It would weakly broadcast sound into the surrounding air, but all four of the lowest resonant frequencies are below the normal human hearing range (20 to 17 000 Hz), so the sounds are not even audible.

P14.23   A string on a cello vibrates in its first normal mode with a frequency of 220 Hz. The vibrating segment is 70.0 cm long and has a mass of 1.20 g.

(a)   Find the tension in the string.

(b)   Determine the frequency of vibration when the string vibrates in three segments.

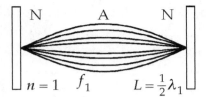

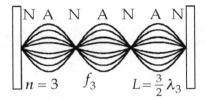

FIG. P14.23

Solution   **Conceptualize:** The tension should be less than 500 N (~100 lb) since excessive force on the four cello strings would break the neck of the instrument. If a string vibrates in three segments, there will be three antinodes (instead of one for the fundamental mode), so the frequency should be three times greater than the fundamental.

Categorize: From the string's length, we can find the wavelength. We then can use the wavelength with the fundamental frequency to find the wave speed. Finally, we can find the tension from the wave speed and the linear mass density of the string.

Analyze: When the string vibrates in the lowest frequency mode, the length of string forms a standing wave where $L = \dfrac{\lambda}{2}$ so the fundamental harmonic wavelength is

$$\lambda = 2L = 2(0.700 \text{ m}) = 1.40 \text{ m}$$

and the speed is

$$v = f\lambda = \left(220 \text{ s}^{-1}\right)(1.40 \text{ m}) = 308 \text{ m/s}.$$

(a)   From the tension equation, $v = \sqrt{\dfrac{T}{\mu}} = \sqrt{\dfrac{T}{m/L}}$.

We get $T = \dfrac{v^2 m}{L}$, or $T = \dfrac{v^2 m}{L} = \dfrac{(308 \text{ m/s})^2 \left(1.20 \times 10^{-3} \text{ kg}\right)}{0.700 \text{ m}} = 163 \text{ N}$.   ◊

continued on next page

(b) For the third harmonic, the tension, linear density, and speed are the same, but the string vibrates in three segments. Thus, that the wavelength is one third as long as in the fundamental.

$$\lambda_3 = \frac{\lambda}{3}$$

From the equation $v = f\lambda$, we find that the frequency is three times as high:

$$f_3 = \frac{v}{\lambda_3} = 3\frac{v}{\lambda} = 3f = 660 \text{ Hz}$$ ◊

Finalize: The tension seems reasonable, and the third harmonic is three times the fundamental frequency as expected. Related to part (b), some stringed instrument players use a technique to double the frequency of a note by "cutting" a vibrating string in half. When the string is lightly touched at its midpoint to form a node, the second harmonic is formed, and the resulting note is one octave higher (twice the original fundamental frequency).

P14.27 Calculate the length of a pipe that has a fundamental frequency of 240 Hz if the pipe is

(a) closed at one end.

(b) open at both ends.

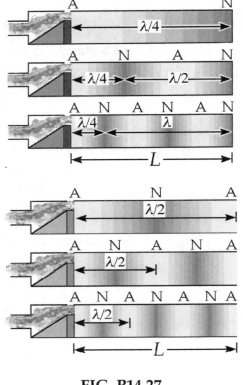

FIG. P14.27

Solution The relationship between the frequency and the wavelength of a sound wave is

$$v = f\lambda \qquad \text{or} \qquad \lambda = \frac{v}{f}.$$

Next, we draw the pipes to help us visualize the relationship between λ and L.

(a) For the fundamental mode in a closed pipe, $\lambda = 4L$

continued on next page

$$\text{so } L = \frac{\lambda}{4} = \frac{v/f}{4} \text{ and } L = \frac{(343 \text{ m/s})/(240 \text{ s}^{-1})}{4} = 0.357 \text{ m}$$ ◊

(b)   For the fundamental mode in an open pipe, $\lambda = 2L$

$$\text{so } L = \frac{\lambda}{2} = \frac{v/f}{2} \text{ and } L = \frac{(343 \text{ m/s})/(240 \text{ s}^{-1})}{2} = 0.715 \text{ m}$$ ◊

P14.31   A shower stall measures 86.0 cm × 86.0 cm × 210 cm. If you were singing in this shower, which frequencies would sound the richest (because of resonance)? Assume that the stall acts as a pipe closed at both ends, with nodes at opposite sides. Assume that the voices of the various singers range from 130 Hz to 2 000 Hz. Let the speed of sound in the hot shower stall be 355 m/s.

Solution   For a closed box, the resonant frequencies will have nodes at both sides, so the permitted wavelengths will be defined by

$$L = \frac{n\lambda}{2} = \frac{nv}{2f}, \quad \text{with } n = 1, 2, 3, \dots.$$

Rearranging, and substituting $L = 0.860$ m, the side-to-side resonant frequencies are

$$f_n = n\frac{v}{2L} = n\frac{355 \text{ m/s}}{2(0.860 \text{ m})} = n(206 \text{ Hz}), \quad \text{for each } n \text{ from 1 to 9 .}$$ ◊

With $L' = 2.10$ m, the top-to-bottom resonance frequencies are

$$f_n = n\frac{355 \text{ m/s}}{2(2.10 \text{ m})} = n(84.5 \text{ Hz}), \quad \text{for each } n \text{ from 2 to 23 .}$$ ◊

P14.33   If two adjacent natural frequencies of an organ pipe are determined to be 550 Hz and 650 Hz, calculate the fundamental frequency and length of this pipe. (Use $v = 340$ m/s.)

continued on next page

Solution We use the wave under boundary conditions model.

Because harmonic frequencies are given by $f_1 n$ for open pipes, and $f_1 = (2n-1)$ for closed pipes, the difference between all adjacent harmonics is constant. Therefore, we can find each harmonic below 650 Hz by subtracting

$$\Delta f_{\text{Harmonic}} = (650 \text{ Hz} - 550 \text{ Hz}) = 100 \text{ Hz from the previous value.}$$

The harmonics are:

$$\{650 \text{ Hz}, 550 \text{ Hz}, 450 \text{ Hz}, 350 \text{ Hz}, 250 \text{ Hz}, 150 \text{ Hz, and } 50 \text{ Hz}\}$$

(a)   The fundamental frequency, then, is 50 Hz.   ◊

(b)   The wavelength of the fundamental frequency can be calculated from the velocity:

$$\lambda = \frac{v}{f} = \frac{340 \text{ m/s}}{50.0 \text{ Hz}} = 6.80 \text{ m}.$$

Because the step size Δf is twice the fundamental frequency, we know the pipe is closed, with an antinode at the open end, and a node at the closed end. The wavelength in this situation is four times the pipe length, so

$$L = \frac{\lambda}{4} = 1.70 \text{ m}.$$ ◊

P14.37   An air column in a glass tube is open at one end and closed at the other by a movable piston. The air in the tube is warmed above room temperature, and a 384-Hz tuning fork is held at the open end. Resonance is heard when the piston is 22.8 cm from the open end and again when it is 68.3 cm from the open end.

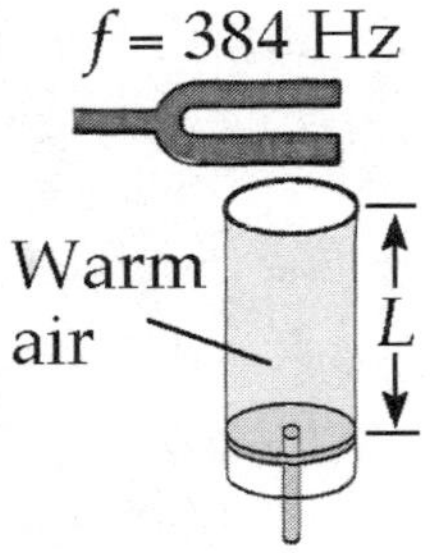

$f = 384$ Hz

Warm
air

FIG. P14.37

(a)   What speed of sound is implied by these data?

(b)   How far from the open end will the piston be when the next resonance is heard?

continued on next page

Solution For an air column closed at one end, resonances will occur when the length of the column is equal to $\dfrac{\lambda}{4}$, $\dfrac{3\lambda}{4}$, $\dfrac{5\lambda}{4}$, and so on. Thus, the change in the length of the pipe from one resonance to the next is $\dfrac{\lambda}{2}$. In this case,

$$\frac{\lambda}{2} = (0.683 \text{ m} - 0.228 \text{ m}) = 0.455 \text{ m and } \lambda = 0.910 \text{ m.}$$

(a) $v = f\lambda = (384 \text{ Hz})(0.910 \text{ m}) = 349 \text{ m/s}$ ◊

(b) $L = 0.683 \text{ m} + 0.455 \text{ m} = 1.14 \text{ m}$ ◊

P14.39   In certain ranges of a piano keyboard, more than one string is tuned to the same note to provide extra loudness. For example, the note at 110 Hz has two strings at this frequency. If one string slips from its normal tension of 600 N to 540 N, what beat frequency is heard when the hammer strikes two strings simultaneously?

Solution **Conceptualize:** Directly noticeable beat frequencies are usually only a few Hertz, so we should not expect a frequency much greater than this.

Categorize: As in previous problems, the two wave speed equations can be used together to find the frequency of vibration that corresponds to a certain tension. The beat frequency is then just the difference in the two resulting frequencies from the two strings with different tensions.

Analyze: Combining the velocity equation $v = f\lambda$ and the tension equation

$$v = \sqrt{\frac{T}{\mu}},$$

we find that $f = \sqrt{\dfrac{T}{\mu\lambda^2}}$.

Since μ and λ are constant, we can apply that equation to both frequencies, and then divide the two equations to get

$$\frac{f_1}{f_2} = \sqrt{\frac{T_1}{T_2}}.$$

With $f_1 = 110$ Hz, $T_1 = 600$ N, and $T_2 = 540$ N:

continued on next page

$$f_2 = (110 \text{ Hz})\sqrt{\frac{540 \text{ N}}{600 \text{ N}}} = 104.4 \text{ Hz}$$

The beat frequency is: $f_b = |f_1 - f_2| = 110 \text{ Hz} - 104.36 \text{ Hz} = 5.64 \text{ Hz}$.   ◊

Finalize: As expected, the beat frequency is only a few cycles per second. This result from the interference of the two sound waves with slightly different frequencies has a tone that varies in amplitude over time, similar to the sound made by saying "wa-wa-wa…"

Note: The beat frequency above is written with three significant figures on the assumption that the original data is known precisely enough to warrant them. This assumption implies that the original frequency is known somewhat more precisely than to three significant digits quoted in "110 Hz." For example, if the original frequency of the strings were 109.6 Hz, the beat frequency would be 5.62 Hz.

P14.41   A student holds a tuning fork oscillating at 256 Hz. He walks toward a wall at a constant speed of 1.33 m/s.

(a)   What beat frequency does he observe between the tuning fork and its echo?

(b)   How fast must he walk away from the wall to observe a beat frequency of 5.00 Hz?

Solution   For an echo, $f_1 = f_2 \dfrac{v + v_s}{v - v_s}$

and the beat frequency $f_b = |f_1 - f_2|$.

Solving for f_b gives $f_b = f_2 \dfrac{2v_s}{v - v_s}$ when approaching the wall.

(a) $f_b = (256 \text{ Hz})\dfrac{2(1.33 \text{ m/s})}{343 \text{ m/s} - 1.33 \text{ m/s}} = 1.99 \text{ Hz}$ ◊

(b)   When moving away from wall, v_s changes sign. Solving for v_s gives

$$v_s = f_b \frac{v}{2f_2 - f_b} = (5.00 \text{ Hz})\frac{343 \text{ m/s}}{2(256 \text{ Hz}) - 5.00 \text{ Hz}} = 3.38 \text{ m/s}.$$ ◊

P14.55   A standing wave is set up in a string of variable length and tension by a vibrator of variable frequency. Both ends of the string are fixed. When the vibrator has a frequency f in a string of length L and tension T, n antinodes are set up in the string.

(a)   If the length of the string is doubled, by what factor should the frequency be changed so that the same number of antinodes is produced?

(b)   If the frequency is tripled and the length are held constant, what tension will produce $n+1$ antinodes?

(c)   If the frequency is tripled and the length of the string is halved, by what factor should the tension be changed so that twice as many antinodes are produced?

Solution   Combining the equations $v = f\lambda$ and $v = \sqrt{\dfrac{T}{\mu}}$ and noting that $\lambda_n = \dfrac{2L}{n}$,

(a)   we find that
$$f_n = \frac{n}{2L}\sqrt{\frac{T}{\mu}} \qquad (1)$$

Keeping n, T, and μ constant, we can create two equations:

$$f_n L = \frac{n}{2}\sqrt{\frac{T}{\mu}}$$

and $f_n' L' = \dfrac{n}{2}\sqrt{\dfrac{T}{\mu}}$.

Dividing the equations, $\dfrac{f_n}{f_n'} = \dfrac{L'}{L}$.

If $L' = 2L$, then $f_n' = \dfrac{1}{2} f_n$.

Therefore, in order to double the length but keep the same number of antinodes, the frequency should be halved.   ◊

(b)   From the same equation (1), we can hold L and f_n constant to get

$$\frac{n'}{n} = \sqrt{\frac{T}{T'}}.$$

continued on next page

From this relation, we see that the tension must be decreased to

$$T' = T\left(\frac{n}{n+1}\right)^2 \text{ to produce } n+1 \text{ antinodes.}$$ ◊

(c)   This time, we rearrange equation (1) to produce

$$\frac{2f_nL}{n} = \sqrt{\frac{T}{\mu}}$$

and $$\frac{2f_n'L'}{n'} = \sqrt{\frac{T'}{\mu}}$$

$$\frac{T'}{T} = \left(\frac{f_n'}{f_n} \cdot \frac{n}{n'} \cdot \frac{L'}{L}\right)^2 = \left(\frac{3f_n}{f_n}\right)^2\left(\frac{n}{2n}\right)^2\left(\frac{L/2}{L}\right)^2 = \frac{9}{16}.$$ ◊

Chapter 15

Fluid Mechanics

Section 15.2   Variation of Pressure with Depth

Section 15.3   Pressure Measurements

The absolute pressure of a fluid is the sum of the gauge pressure and atmospheric pressure. The SI unit of pressure is the Pascal (Pa). Note that $1 \text{ Pa} \equiv 1 \text{ N/m}^2$. *In a fluid at rest, all points at the same depth are at the same pressure.*

Pascal's law states that a change in pressure applied to an enclosed fluid is transmitted undiminished to every point in the fluid and the walls of the containing vessel.

Section 15.4   Buoyant Forces and Archimedes's Principle

Any object partially or completely submerged in a fluid experiences a buoyant force equal in magnitude to the weight of the fluid displaced by the object and acting vertically upward through the point which was the center of gravity of the displaced fluid.

Section 15.5   Fluid Dynamics

When fluid is in motion, its flow can be characterized as being one of two main types. The flow is said to be steady if each particle of the fluid follows a smooth path, and the paths of the particles do not cross. Above a certain critical speed, fluid flow becomes nonsteady or turbulent, characterized by small whirlpool-like regions.

The term viscosity is commonly used in fluid flow to characterize the degree of internal friction in the fluid. This internal friction is associated with the resistance to two adjacent layers of the fluid to move relative to each other. Because of viscosity, part of the kinetic energy of a fluid is converted to internal energy.

In our model of an ideal fluid, we make the following four assumptions:

- **Nonviscous fluid**. In a nonviscous fluid, internal friction is neglected. An object moving through a nonviscous fluid would experience no retarding viscous force.

- **Steady flow**. In a steady flow, we assume that the velocity of the fluid at each point remains constant in time.

- **Incompressible fluid**. The density of the fluid is assumed to remain constant in time.

- **Irrotational flow**. Fluid flow is irrotational if there is no angular momentum of the fluid about any point. If a small wheel placed anywhere in the fluid does not rotate about its center of mass, the flow would be considered irrotational. (If the wheel were to rotate, as it would if turbulence were present, the flow would be rotational.)

Fluids that have the "ideal" properties stated above obey two important equations:

- The **equation of continuity** states that the flow rate through a pipe is constant (i.e. the product of the cross-sectional area of the pipe and the speed of the fluid is constant).

- **Bernoulli's equation** states that the sum of the pressure (P), kinetic energy per unit volume $\left(\dfrac{\rho v^2}{2} \right)$, and the potential energy per unit volume ($\rho g h$) has a constant value at all points along a streamline.

Section 15.6   Streamlines and the Continuity Equation for Fluids

The path taken by a fluid particle under steady flow is called a streamline. The velocity of the fluid particle is always tangent to the streamline at that point and no two streamlines can cross each other. A set of streamlines forms a tube of flow. In steady flow, particles of the fluid cannot flow into or out of a tube of flow.

EQUATIONS AND CONCEPTS

The **density of a homogeneous substance** is defined as its ratio of mass to volume. The value of density is characteristic of a particular type of material and independent of the total quantity of material in the sample. The **SI units of density** are kg per cubic meter.

$$\rho \equiv \frac{m}{V}$$

$$1 \text{ g/cm}^3 = 1000 \text{ kg/m}^3$$

The **(average) pressure of a fluid** is defined as the normal force per unit area acting on a surface immersed in the fluid. *Pressure is a scalar quantity.* The SI units of pressure are newtons per square meter, or Pascal (Pa).

$$P \equiv \frac{F}{A} \tag{15.1}$$

$$1 \text{ Pa} \equiv 1 \text{ N/m}^2 \tag{15.2}$$

Atmospheric pressure is often expressed in other units: atmospheres, mm of mercury (Torr), or pounds per square inch.

$$P_0 = 1 \text{ atm} \approx 1.013 \times 10^5 \text{ Pa} \tag{15.3}$$

$$1 \text{ Torr} = 133.3 \text{ Pa}$$

$$1 \text{ lb/in}^2 = 6.89 \times 10^3 \text{ Pa}$$

The **absolute pressure**, P, at a depth, h, below the surface of a liquid which is open to the atmosphere is greater than atmospheric pressure, P_0, by an amount which depends on the depth below the surface.

$$P = P_0 + \rho g h \tag{15.4}$$

$$P_{\text{absolute}} = P_{\text{atmosphere}} + P_{\text{gauge}}$$

The pressure has the same value at all points at a given depth and does not depend on the shape of the container.

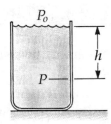

Pascal's law states that pressure applied to an enclosed fluid (liquid or gas) is transmitted undiminished to every point within the fluid and to the walls of the vessel which contain the fluid.

Pascal's law

Archimedes's principle states that when an object is partially or fully immersed in a fluid, the fluid exerts an upward buoyant force on the object. The magnitude of the buoyant force equals the weight of the fluid displaced by the object. The weight of the displaced fluid depends on the fluid density, and the volume of fluid displaced.

$$B = \rho_f g V \tag{15.5}$$

(Archimedes's principle)

For a **floating object**, the fraction of the object below the fluid surface equals the ratio of the object density to the fluid density.

$$\frac{\rho_0}{\rho_f} = \frac{V}{V_0} \tag{15.6}$$

V = volume submerged
V_0 = volume of object

The **equation of continuity** for an incompressible fluid (ρ = constant), states that the flow rate at any point along a pipe carrying an incompressible fluid is constant.

$$A_1 v_1 = A_2 v_2 \tag{15.7}$$

Av = volume flow rate

Bernoulli's equation is a fundamental law in fluid mechanics. The equation is a statement of the law of conservation of mechanical energy as applied to a fluid. Bernoulli's equation states that the sum of pressure, kinetic energy per unit volume, and gravitational potential energy per unit volume remains constant throughout an ideal fluid.

$$P + \frac{1}{2}\rho v^2 + \rho g y = \text{constant} \tag{15.13}$$

(Bernoulli's equation)

REVIEW CHECKLIST

✓   Understand the concept of pressure at a point in a fluid and the variation of pressure with depth.

✓   Understand the relationships among absolute, gauge, and atmospheric pressure values; and know the several different units commonly used to express pressure.

✓   Understand the origin of buoyant forces; and state and explain Archimedes's principle.

✓   State the simplifying assumptions of an ideal fluid moving with streamline flow.

✓   State and understand the physical significance of the equation of continuity (constant flow rate) and Bernoulli's equation for fluid flow (relating flow velocity, pressure, and pipe elevation).

ANSWERS TO SELECTED QUESTIONS

Q15.1   Two drinking glasses having equal weights but different shapes and different cross-sectional areas are filled to the same level with water. According to the expression $P = P_0 + \rho g h$, the pressure is the same at the bottom of both glasses. In view of this fact, why does one weigh more than the other?

Answer   For the cylindrical container shown, the weight of the water is equal to the gauge pressure at the bottom multiplied by the area of the bottom. For the narrow-necked bottle, on the other hand, the weight of the fluid is much less than PA, the bottom pressure times the area. The water does exert the large PA force downward on the bottom of the bottle, but the water also exerts an upward force nearly as large on the ring-shaped horizontal area surrounding the neck. The net vector force that the water exerts on the glass is equal to the small weight of the water.

FIG. Q15.1

Q15.5   A fish rests on the bottom of a bucket of water while the bucket is being weighed on a scale. When the fish begins to swim around, does the scale reading change?

Answer   In either case, the scale is supporting the container, the water, and the fish. Therefore the weight remains the same. The reading on the scale, however, can change if the net center of mass accelerates in the vertical direction, as when the fish jumps out of the water. In that case, the scale, which registers force, will also show an additional force caused by Newton's law, $F = ma$.

FIG. Q15.5

Q15.9   The water supply for a city is often provided from reservoirs built on high ground. Water flows from the reservoir, through pipes, and into your home when you turn the tap on your faucet. Why is the water flow more rapid out of a faucet on the first floor of a building than in an apartment on a higher floor?

Answer   The water supplied to the building flows through a pipe connected to the water tower. Near the earth, the water pressure is greater because the pressure increases with increasing depth beneath the surface of the water. The penthouse apartment is not as far below the water surface, hence the water flow will not be as rapid as in a lower floor.

Q15.13   A barge is carrying a load of gravel along a river. As it approaches a low bridge, the captain realizes that the top of the pile of gravel is not going to make it under the bridge. The captain orders the crew to shovel gravel quickly from the pile into the water. Is this a good decision?

Answer   It turns out that this is not a good decision. Imagine a given shovel full of gravel. According to Archimedes's principle, the portion of the total buoyant force which can be associated with keeping this shovel-full of gravel afloat on the barge is due to a portion of water whose weight is equal to the weight of the gravel. Since water is less dense than gravel, the volume of water associated with keeping a shovel-full of gravel afloat is larger than the volume of the gravel. When the gravel is thrown overboard, the barge rises, in association with a volume of water larger than the volume of the gravel removed from the pile. Thus, as the gravel is shoveled overboard, the barge rises by an amount which is larger than the amount by which the height of the pile of gravel decreases. The better approach would be to **add** gravel to the barge from a riverside source, causing it to sink deeper into the water at a greater rate than the height of the pile of gravel increases.

Q15.19   An unopened can of diet cola floats when placed in a tank of water, whereas a can of regular cola of the same brand sinks in the tank. What do you suppose could explain this behavior?

Answer   Regular cola is sugar syrup. Its density is higher than the density of diet cola, which is nearly pure water. The low-density air inside the can has a bigger effect than the thin aluminum shell, so the can of diet soda floats.

SOLUTIONS TO SELECTED PROBLEMS

P15.1   Calculate the mass of a solid iron sphere that has a diameter of 3.00 cm.

Solution   The definition of density $\rho = \dfrac{m}{V}$ is often written as $m = \rho V$.

Here $V = \dfrac{4}{3}\pi r^3$ so $m = \rho V = \rho\left(\dfrac{4}{3}\pi\left(\dfrac{d}{2}\right)^3\right)$.

Thus, $m = \left(7.86 \times 10^3 \ \text{kg/m}^3\right)\left(\dfrac{4}{3}\right)\pi(1.50 \ \text{cm})^3\left(10^{-6} \ \text{m}^3/\text{cm}^3\right)\left(10^3 \ \text{g/kg}\right) = 111 \ \text{g}.$ ◊

P15.3   A 50.0-kg woman balances on one heel of a pair of high-heeled shoes. If the heel is circular and has a radius of 0.500 cm, what pressure does she exert on the floor?

Solution The area of the circular base of the heel is

$$\pi r^2 = \pi(0.500 \text{ cm})^2 \left(\frac{1 \text{ m}^2}{10\ 000 \text{ cm}^2} \right) = 7.85 \times 10^{-5} \text{ m}^2$$

The force she exerts is her weight,

$$mg = (50.0 \text{ kg})(9.80 \text{ m/s}^2) = 490 \text{ N}.$$

Then $P = \dfrac{F}{A} = \dfrac{490 \text{ N}}{7.85 \times 10^{-5} \text{ m}^2} = 6.24 \text{ MPa}.$ ◊

P15.5   The spring of the pressure gauge shown in Figure 15.5 has a force constant of 1 000 N/m and the piston has a diameter of 2.00 cm. As the gauge is lowered into water, what change in depth causes the piston to move in by 0.500 cm?

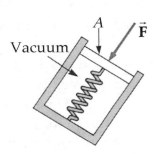

Solution $F_{\text{spring}} = F_{\text{fluid}}$ or $kx = \rho g h A$

FIG. 15.5

$$h = \frac{kx}{\rho g A} = \frac{(1\ 000 \text{ N/m}^2)(0.005\ 00 \text{ m})}{(1\ 000 \text{ kg/m}^3)(9.80 \text{ m/s}^2)(0.010\ 0 \text{ m})^2 \pi} = 1.62 \text{ m}$$ ◊

P15.7   What must be the contact area between a suction cup (completely exhausted) and a ceiling if the cup is to support the weight of an 80.0-kg student?

Solution **Conceptualize:** The suction cups used by burglars seen in movies are about 10 cm in diameter, and it seems reasonable that one of these might be able to support the weight of an 80-kg student. The face area of a 10-cm cup is approximately:

$$A = \pi r^2 \approx 3(0.05 \text{ m})^2 \approx 0.008 \text{ m}^2$$

FIG. 15.7

continued on next page

Categorize: "Suction" is not a new kind of force. Familiar forces hold the cup in equilibrium, one of which is the atmospheric pressure acting over the area of the cup. This problem is simply another application of Newton's 2nd law.

Analyze: The vacuum between cup and ceiling exerts no force on either. The atmospheric pressure of the air below the cup pushes up on it with a force $(P_{atm})(A)$. If the cup barely supports the student's weight, then the normal force of the ceiling is approximately zero, and

$$\sum F_y = 0 + (P_{atm})(A) - mg = 0: \ A = \frac{mg}{P_{atm}} = \frac{784 \text{ N}}{1.013 \times 10^5 \text{ N/m}^2} = 7.74 \times 10^{-3} \text{ m}^2 \qquad \Diamond$$

Finalize: This calculated area agrees with our prediction and corresponds to a suction cup that is 9.93 cm in diameter. (Our 10 cm estimate was right on—a lucky guess, considering that a burglar would probably use at least two suction cups, not one.) As an aside, the suction cup we have drawn, above, appears to be about 30 cm in diameter, plenty big enough to support the weight of the student.

P15.13   Blaise Pascal duplicated Torricelli's barometer using a red Bordeaux wine of density 984 kg/m^3 as the working liquid (Fig. P15.13). What was the height h of the wine column for normal atmospheric pressure? Would you expect the vacuum above the column to be as good as for mercury?

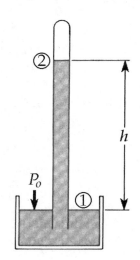

FIG. P15.13

Solution   In Bernoulli's equation,

$P_1 + \dfrac{1}{2}\rho v_1^2 + \rho g y_1 = P_2 + \dfrac{1}{2}\rho v_2^2 + \rho g y_2$. Take point 1 at the wine surface in the pan, where $P_1 = P_{atm}$, and point 2 at the wine surface up in the tube. Here we approximate $P_2 = 0$, although some alcohol and water will evaporate. The vacuum is not as good as with mercury. Unless you are careful, a lot of dissolved oxygen or carbon dioxide may come bubbling out.

   Now, since the speed of the fluid at both points is zero, we can rearrange the terms to arrive essentially at Equation 15.4:

$$P_1 = P_2 + \rho g(y_2 - y_1)$$
$$1 \text{ atm} = 0 + (984 \text{ kg/m}^3)(9.80 \text{ m/s}^2)(y_2 - y_1)$$
$$y_2 - y_1 = \frac{1.013 \times 10^5 \text{ N/m}^2}{9\,643 \text{ N/m}^3} = 10.5 \text{ m} \qquad \Diamond$$

A water barometer in a stairway of a three-story building is a nice display. Red wine makes the fluid level easier to see.

P15.19 A Ping-Pong ball has a diameter of 3.80 cm and average density of $0.084\,0$ g/cm^3. What force is required to hold it completely submerged under water?

Solution **Conceptualize:** According to Archimedes's Principle, the buoyant force acting on the submerged ball will be equal to the weight of the water the ball will displace. The ball has a volume of about 30 cm^3, so the weight of this water is approximately:

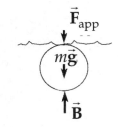

$$B = F_g = \rho V g \approx \left(1\ \text{g/cm}^3\right)\left(30\ \text{cm}^3\right)\left(10\ \text{m/s}^2\right) = 0.3\ \text{N}$$

FIG. P15.19

Since the ball is much less dense than the water, the applied force will approximately equal this buoyant force.

Categorize: Apply Newton's 2nd law to find the applied force.

Analyze: At equilibrium, $\sum F = 0$ or $-F_{\text{app}} - mg + B = 0$

where the buoyant force is $B = \rho_w V g$ and $\rho_w = 1\,000$ kg/m^3.

The applied force is then $F_{\text{app}} = \rho_w V g - mg$.

Using $m = \rho_{\text{ball}} V$ to eliminate the unknown mass of the ball, this becomes

$$F_{\text{app}} = V g (\rho_w - \rho_{\text{ball}}) = \frac{4}{3} \pi r^3 g (\rho_w - \rho_{\text{ball}})$$

$$F_{\text{app}} = \frac{4}{3} \pi (1.90 \times 10^{-2}\ \text{m})^3 (9.80\ \text{m/s}^2)(1\,000\ \text{kg/m}^3 - 84\ \text{kg/m}^3)$$

$$F_{\text{app}} = 0.258\ \text{N} \qquad \lozenge$$

Finalize: The force is approximately what we expected, so our result is reasonable. If a force is greater than 0.258 N were to be applied, the ball would accelerate down until it hit the bottom (which would then provide a normal force directed upwards).

P15.23 A cube of wood having an edge dimension of 20.0 cm and a density of 650 kg/m^3 floats on water.

(a) What is the distance from the horizontal top surface of the cube to the water level?

(b) What mass of lead weight should be placed on top of the cube so that its top is just level with the water?

continued on next page

Solution Set h equal to the distance from the top of the cube to the water level.

(a)   According to Archimedes's principle,

$$B = \rho_w V g = \left(1.00 \text{ g/cm}^3\right)\left[(20.0 \text{ cm})^2(20.0 \text{ cm} - h \text{ cm})\right]g.$$

But B = Weight of block = $mg = \rho_{\text{wood}} V_{\text{wood}} g = \left(0.650 \text{ g/cm}^3\right)(20.0 \text{ cm})^3 g$.

Setting these two equations equal,

$$\left(0.650 \text{ g/cm}^3\right)(20.0 \text{ cm})^3 g = \left(1.00 \text{ g/cm}^3\right)(20.0 \text{ cm})^2(20.0 \text{ cm} - h \text{ cm})g$$
$$20.0 \text{ cm} - h = 20.0(0.650) \text{ cm}$$
$$h = 20.0(1.00 \text{ cm} - 0.650 \text{ cm}) = 7.00 \text{ cm} \qquad \Diamond$$

(b) $B = mg + Mg$ where M = mass of lead

$$\left(1.00 \text{ g/cm}^3\right)(20.0 \text{ cm})^3 g = \left(0.650 \text{ g/cm}^3\right)(20.0 \text{ cm})^3 g + Mg$$
$$M = (20.0 \text{ cm})^3\left(1.00 \text{ g/cm}^3 - 0.650 \text{ g/cm}^3\right) = (20.0 \text{ cm})^3\left(0.350 \text{ g/cm}^3\right)$$
$$M = 2.80 \text{ kg} \qquad \Diamond$$

P15.25   How many cubic meters of helium are required to lift a balloon with a 400-kg payload to a height of 8 000 m? (Take $\rho_{\text{He}} = 0.180 \text{ kg/m}^3$.) Assume that the balloon maintains a constant volume and that the density of air decreases with altitude z according to the expression $\rho_{\text{air}} = \rho_0 e^{-z/8\,000}$, where z is in meters, and $\rho_0 = 1.25 \text{ kg/m}^3$ is the density of air at sea level.

Solution   At $z = 8\,000$ m, the density of air is

$$\rho_{\text{air}} = \rho_0 e^{-z/8\,000} = \left(1.25 \text{ kg/m}^3\right)e^{-1} = \left(1.25 \text{ kg/m}^3\right)(0.368) = 0.460 \text{ kg/m}^3$$

Think of the balloon reaching equilibrium at this height.

The weight of its payload is $Mg = (400 \text{ kg})(9.80 \text{ m/s}^2) = 3\,920 \text{ N}$.

The weight of the helium in it is $mg = \rho_{\text{He}} V g$.

$\sum F_y = 0$ becomes $+\rho_{\text{air}} V g - Mg - \rho_{\text{He}} V g = 0$.

Solving, $(\rho_{\text{air}} - \rho_{\text{He}})V = M$

and $V = \dfrac{M}{\rho_{\text{air}} - \rho_{\text{He}}} = \dfrac{400 \text{ kg}}{(0.460 - 0.18) \text{ kg/m}^3} = 1.43 \times 10^3 \text{ m}^3.$ $\qquad \Diamond$

P15.27 A plastic sphere floats in water with 50.0% of its volume submerged. This same sphere floats in glycerin with 40.0% of its volume submerged. Determine the densities of the glycerin and the sphere.

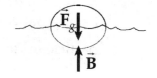

Solution The forces on the ball are its weight

$$F_g = mg = \rho_{\text{plastic}} V_{\text{ball}} g$$

FIG. P15.27

and the buoyant force of the liquid $B = \rho_{\text{fluid}} V_{\text{immersed}} g$.

When floating in water, $\sum F_y = 0$:

$$-\left(\rho_{\text{plastic}}\right)\left(V_{\text{ball}}\right)g + \left(\rho_{\text{water}}\right)\left(0.500 V_{\text{ball}}\right)g = 0$$

$$\rho_{\text{plastic}} = 0.500 \rho_{\text{water}} = 500 \text{ kg/m}^3 \qquad\qquad \Diamond$$

When floating in glycerin, $\sum F_y = 0$:

$$-\left(\rho_{\text{plastic}}\right)\left(V_{\text{ball}}\right)g + \left(\rho_{\text{glycerin}}\right)\left(0.400 V_{\text{ball}}\right)g = 0$$

$$\rho_{\text{plastic}} = 0.400 \rho_{\text{glycerin}}$$

$$\rho_{\text{glycerin}} = \frac{500 \text{ kg/m}^3}{0.400} = 1\,250 \text{ kg/m}^3 \qquad\qquad \Diamond$$

This glycerin would sink in water.

P15.33 A large storage tank with an open top is filled to a height h_0. The tank is punctured at a height h above the bottom of the tank (Fig. P15.33). Find an expression for how far from the tank the exiting stream lands.

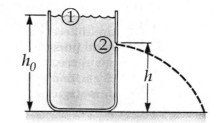

Solution Take point 1 at the top liquid surface. Since the tank is large, the fluid level falls only slowly and $v_1 \approx 0$. Take point 2 at the surface of the water stream leaving the hole. At both points the pressure is one atmosphere because the water can push no more or less strongly than the air pushes on it, as described by Newton's third law.

FIG. P15.33

$$P_1 + \frac{1}{2}\rho v_1^2 + \rho g y_1 = P_2 + \frac{1}{2}\rho v_2^2 + \rho g y_2$$

$$P_a + 0 + \rho g h_0 = P_a + \frac{1}{2}\rho v_2^2 + \rho g h$$

$$v_2 = \sqrt{2g(h_0 - h)}$$

continued on next page

Now each drop of water moves as a projectile. Its velocity v_2, which we now call its original velocity, has zero vertical component. Its time of fall is given by the particle under constant acceleration model, from $y = v_{y0}t + \frac{1}{2}a_y t^2$:

$$-h = 0 - \frac{1}{2}gt^2 \qquad \text{so} \qquad t = \sqrt{\frac{2h}{g}}$$

and from the particle under constant velocity model its horizontal displacement is

$$x = v_{x0}t + \frac{1}{2}a_x t^2 = \sqrt{2g(h_0 - h)}\sqrt{\frac{2h}{g}} + 0 = \sqrt{4h(h_0 - h)} . \qquad \Diamond$$

Note that as h_0 goes to zero the speed of efflux and range approach zero. As h goes to zero the range goes to zero, since water leaking horizontally from the bottom edge of the tank will have no space for a free-fall trajectory. Also, as h approaches h_0 the range goes to zero: water does not spray from the top of the tank.

P15.49   The true weight of an object is measured in a vacuum, where buoyant forces are absent. An object of volume V is weighed in air on an equal-arm balance with the use of counterweights of density ρ. Let the density of air be ρ_{air} and the balance reading F_g'. Show that the true weight F_g is

$$F_g = F_g' + \left(V - \frac{F_g'}{\rho g}\right)\rho_{air} g .$$

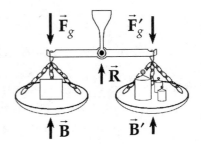

FIG. P15.49

Solution   We use the rigid body in equilibrium model. The "balanced" condition is one in which the net torque on the balance is zero. Since the balance has lever arms of equal length, the total force on each pan is equal. Applying $\sum \tau = 0$ around the pivot gives us this in equation form:

$$F_g - B = F_g' - B'$$

where B and B' are the buoyant forces on the body and weights respectively. The buoyant force experienced by an object of volume V in air is

$$B = V\rho_{air} g .$$

continued on next page

So for the test mass and for the weights, respectively,

$$B = V\rho_{air}g \text{ and } B' = V'\rho_{air}g.$$

Since the volume of the weights is not given explicitly, we must use the density equation to eliminate it:

$$V' = \frac{m'}{\rho} = \frac{m'g}{\rho g} = \frac{F'_g}{\rho g}.$$

With this substitution, the buoyant force on the weights is $B' = \left(\dfrac{F'_g}{\rho g}\right)\rho_{air}g.$

Therefore, $F_g = F'_g + \left(V - \dfrac{F'_g}{\rho g}\right)\rho_{air}g.$ ◊

Related Comment: We can now answer the popular riddle: Which weighs more, a pound of feathers or a pound of bricks? Like in the problem above, the feathers have a greater buoyant force than the bricks, so if they "weigh" the same on a scale as a pound of bricks, then the feathers must have more mass and therefore a greater "true weight."

P15.53 **Review problem.** With reference to Figure 15.53, show that the total torque exerted by the water behind the dam about a horizontal axis through O is $\dfrac{1}{6}\rho gwH^3$. Show that the effective line of action of the total force exerted by the water is at a distance $\dfrac{1}{3}H$ above O.

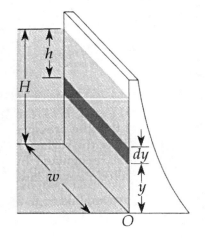

FIG. 15.53

Solution The torque is calculated from the equation
$$\tau = \int d\tau = \int r \, dF.$$

From Figure 15.53, we have

$$\tau = \int (y) \, dF$$

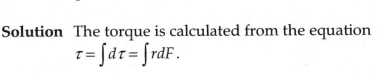

$$\tau = \int_0^H y\left[\rho g(H-y)w\right]dy = \rho g H w \int_0^H y \, dy - \rho g w \int_0^H y^2 \, dy = \rho g H w \frac{y^2}{2}\bigg|_0^H - \rho g w \frac{y^3}{3}\bigg|_0^H$$

$$= \rho g H w \frac{H^2}{2} - \rho g w \frac{H^3}{3} = \frac{1}{6}\rho g w H^3 \qquad ◊$$

continued on next page

Similarly for the force exerted by the water,

$$F = \int dF = \int_0^H (\rho g (H - y) w) dy = \rho g H w \int_0^H dy - \rho g w \int_0^H y \, dy = \rho g H w H - \rho g w \frac{H^2}{2} = \frac{1}{2} \rho g w H^2.$$

If this were applied at a height y_{eff} such that the torque remains unchanged,

$$\frac{1}{6} \rho g w H^3 = y_{\text{eff}} \left[\frac{1}{2} \rho g w H^2 \right]$$

and

$$y_{\text{eff}} = \frac{1}{3} H. \qquad\qquad \Diamond$$

The direct proportionality of the torque to the density of the fluid, the free-fall acceleration, and the width of the dam are all reasonable. Note that the torque does not depend on the volume of water the dam is holding back. It does not depend on the length of the lake. The proportionality of the torque to the cube of the depth is remarkable. It is required (for one thing) for dimensional correctness of the expression. In units,

$$\left[\frac{1}{6} \rho g w H^3 \right] = \frac{\text{kg}}{\text{m}^3} \frac{\text{m}}{\text{s}^2} \text{m} \cdot \text{m}^3 = \frac{\text{kg} \cdot \text{m}^2}{\text{s}^2} = \text{N} \cdot \text{m}$$

and the unit of torque is indeed $\text{N} \cdot \text{m}$. The water pressure increases with depth. More than half of the force on the dam is exerted on its lower half. This is why the line of action of the net force is below $\frac{1}{2} H$, namely at $\frac{1}{3} H$ above the lake bottom.